Praise for DIY Mushroom Cultivation

DIY Mushroom Cultivation is a valuable resource for a to begin or advance their mushroom cultivation practice. Offering clear and comprehensive instructions for low-tech growing for a range of budgets, interests and scales, this book offers practical inspiration and a sense that: "hey, I can do this!" Willoughby has a refreshingly warm way of communicating about the art and science of mushroom cultivation, which welcomes the reader into a community of delight and appreciation of the fungi and the many ways we can work with them to support human and ecological health and well-being.

DANIELLE STEVENSON owner, DIY Fungi

With hands of an artist, eyes of an ecologist, and the heart of a deeply connected human, Willoughby brings the practical and joyful together through the science and wonder of mycology. This volume on mushroom cultivation is an invitation to read and share in community to grow not only mushrooms, but our true earthly relationships.

NANCE KLEHM Director of Social Ecologies, and author, *The Ground Rules* and *The Soil Keepers: Conversations with Practitioners on the Ground Beneath Our Feet*

Willoughby's unique combination of talent, passion, and experience growing and cooking mushrooms has produced this beautiful and informative contribution. Well done, Mycology Maestro!

ROBERT ROGERS author, *The Fungal Pharmacy*

It is simply a fact that as the horrors of climate change descend upon us, ecosystem regeneration will become our overarching drive. Then, finally, we will seek advice and assistance from our beloved ancestors, the forest fungi. Ever forgiving, they will supply our indoor protein, nurture our new forests, financially support our drawdown effort, and mend our damaged world. This delightful book gets all that started in clear language and beautiful illustrations, at home scale and low budget, one new mushroom grower at a time.

ALBERT BATES owner, Mushroom People, North America's oldest mushroom tools supply company, and author, *The Biochar Solution*.

DIY Mushroom Cultivation by Willoughby Arevalo is a great book for those looking to dig deeper into the powerful world of mushroom cultivation and appreciation. Willoughby's knowledge, passion, and sincere love for fungi makes this book a wonderful, accessible, and informative read. The book provides a solid foundation for inspiring and empowering individuals and communities to grow mushrooms for food, healing, and remediation.

LEILA DARWISH author, *Earth Repair: A Grassroots Guide to Healing Toxic and Damaged Landscapes*

Fun to read and easy to digest, this guide is both practical and inspirational. There are few people in our modern society that have the kind of deep connection with the fungal realm as Willoughby Arevalo—the mushrooms actually speak through him. Both experienced and novice mushroom growers will enjoy the adventure and grow their skills.

MAYA ELSON mycologist and Santa Cruz Program Director, Wild Child Free School

DIY MUSHROOM CULTIVATION

Growing Mushrooms at Home for Food, Medicine, and Soil

WILLOUGHBY AREVALO

ILLUSTRATED BY CARMEN ELISABETH

Copyright © 2019 by Willoughby Arevalo.
All rights reserved.

Cover design by Diane McIntosh.
Cover image: Nameko Mushrooms © Anne Murphy
Back cover images: Upper and Lower Left © Ja Schindler; Right © Willoughby Arevalo.
Interior design by Setareh Ashrafologhalai
Illustrations by Carmen Elisabeth Olson

Printed in Canada. Second printing July 2021.

LIBRARY AND ARCHIVES CANADA CATALOGUING IN PUBLICATION

Title: DIY mushroom cultivation : growing mushrooms at home for food, medicine, and soil / Willoughby Arevalo.

Other titles: Do it yourself mushroom cultivation
Names: Arevalo, Willoughby, 1983- author. | Olson, Carmen Elisabeth, illustrator.
Description: Illustrations by Carmen Elisabeth Olson. | Includes bibliographical references and index.
Identifiers: Canadiana (print) 20190092696 | Canadiana (ebook) 2019009270X | ISBN 9780865718951 (softcover) | ISBN 9781550926880 (PDF) | ISBN 9781771422840 (EPUB)
Subjects: LCSH: Mushroom culture—Handbooks, manuals, etc. | LCSH: Edible mushrooms—Handbooks, manuals, etc. | LCGFT: Handbooks and manuals.
Classification: LCC SB353 .A74 2019 | DDC 635/.8—dc23

This book is intended to be educational and informative. It is not intended to serve as a guide. The author and publisher disclaim all responsibility for any liability, loss or risk that may be associated with the application of any of the contents of this book.

Inquiries regarding requests to reprint all or part of *DIY Mushroom Cultivation* should be addressed to New Society Publishers at the address below. To order directly from the publishers, please call toll-free (North America) 1-800-567-6772, or order online at www.newsociety.com

Any other inquiries can be directed by mail to:
New Society Publishers
P.O. Box 189, Gabriola Island, BC V0R 1X0, Canada
(250) 247-9737

New Society Publishers' mission is to publish books that contribute in fundamental ways to building an ecologically sustainable and just society, and to do so with the least possible impact on the environment, in a manner that models this vision.

DANIELLE STEVENSON

CONTENTS

ACKNOWLEDGMENTS x

INTRODUCTION 1
Mushrooms and Humans: Past, Present, and Future 1
A Bit about Me and My Approach to Mushroom Cultivation 2

1 **MUSHROOM BASICS** 6
What Are Mushrooms? 6
Mushrooms in Ecosystems 13
What Mushrooms Need to Survive and Thrive 17

2 **OVERVIEW OF THE CULTIVATION PROCESS** 20
Cultivation Flow 20
Preparation 24

3 WORKSPACES, TOOLS, AND EQUIPMENT 26
  Lab Infrastructure and Aseptic Transfer Spaces:
    Flow Hoods, Still Air Boxes, and More 26
  The Lab Environment and Tools 30
  Where to Inoculate Bulk Substrates 34
  Where to Incubate Growing Mycelium 34
  Fruiting Space: Factors to Consider 37
  Options for Home-Scale Fruiting Chambers 38
  Environmental Control in Fruiting Spaces 46
  Other Spaces 51

4 SANITATION AND TECHNIQUES TO AVOID CONTAMINATION 54
  Vectors of Contamination and
    Management Strategies 56
  Common Contaminants: Recognition
    and Management 62

5 STARTING AND MAINTAINING CULTURES 66
  Get Cultured 66
  Liquid Culture 69
  Agar Culture 81
  Long-Term Culture Storage Methods 88

6 MAKING AND USING GRAIN SPAWN 90
  Making Grain Spawn 90
  Using Grain Spawn 95

7 **FRUITING SUBSTRATE FORMULATION AND PREPARATION** 98

Containers for Mycelial Growth and Fruiting 98

The Substrates 104

    Substrate Treatments 116

    Sterilization 116

    Pasteurization and Alternatives 118

8 **OUTDOOR GROWING AND MUSHROOM GARDENING** 126

Growing Mushrooms on Logs and Stumps 126

Mushroom Beds 132

Next-Level Applications 135

9 **HARVEST, PROCESSING, AND USE** 138

When and How to Harvest 138

Basic Cooking Techniques 139

Preservation Methods 143

Mushrooms and Mycelium for Medicine 144

IN CONCLUSION: SUBSTRATE FOR THOUGHT—TOWARD FURTHER APPLICATIONS 147

Mycopermaculture 147

Mycoremediation on a Home Scale 150

Mycoarts and Fungi as Functional Materials 150

Community-Based Cultivation Efforts 151

APPENDIX 1: SPECIES PROFILES 153
Agrocybe aegerita— Pioppino 154
Coprinus comatus—Shaggy Mane 155
Cordyceps militaris—Caterpillar Fungus 156
Flammulina velutipes and allies—Enoki 157
Ganoderma lucidum and allies—Reishi 159
Hericium species—Lion's Mane and allies 160
Hypsizygus tessulatus—Shimeji and
    H. ulmarius—Elm Oysters 162
Lentinula edodes—Shiitake 163
Pholiota nameko and allies—Nameko 165
Pleurotus species—Oyster Mushrooms 166
Stropharia rugoso-annulata—Wine Cap 169
Trametes versicolor—Turkey Tail 171

APPENDIX 2: RESOURCES 173
General 173
Annual Mycology Gatherings 177

BIBLIOGRAPHY 178

INDEX 181

ABOUT THE AUTHOR AND ILLUSTRATOR 186

A NOTE ABOUT THE PUBLISHER 188

ACKNOWLEDGMENTS

Homegrown nameko mushrooms. WILLOUGHBY AREVALO

WITH LOVE AND gratitude I recognize the network of heart-forward and hopeful mushroom people who have helped me on my journey and been such good friends, collaborators, and colleagues, including but not limited to: Max Kirchgasser for your companionship and generous help with cultivation projects through my writing process; Mike Egan and Mycality Mushrooms; Max Brotman and Claire Brown; Danielle Stevenson and DIY Fungi; Leila Darwish; Olga Tzogas, Rebar, and The Mycelium Underground/Smugtown Mushrooms; Nance Klehm and The Ground Rules; Maya Elson, Mia Maltz and CoRenewal, Lexie Gropper and Amisacho; Robert Rogers; Ja Schindler, Val Nguyen, and Fungi for the People; Michael Hathaway and the Matsutake Worlds Research Group; Kaitlin Bryson; Nina O'Malley, Charlie Aller, and Mush Luv; Mara Penfil and Female and Fungi; Peter McCoy and Radical Mycology; Jason Leane and All the Mushrooms; Ava Arvest, Raskal Turbeville, and Myco-Uprrhizal; the Mushroom Man Scott Henderson; Octavio Perez Ortiz and Senguihongo; Dallas Lawlor and Northside Fungi Farm; Ray and Patty Lanier and Mushroom Maestros; Willie Crosby and Fungi Ally; Geoffroy Grignon and Champignons Maison/Mycélium Remédium; Philip Zoghibi; Mycollectif; Xiaojing Yan; Ionatan Waisgluss and Vegetation Station;

Theo Rosenfeld and Wildwood Ecology Labs; Nadine Simpson; the online mushroom community; all my cultivation students and teacher's assistants; Paul Kroeger and all my friends and colleagues in the Vancouver Mycological Society; the Humboldt Bay Mycological Society; and my teacher Dr. Terry Henkel and early mentor Dr. David Largent.

To my beloved family for the love and support: Isabelle, Uma, Mom, Dad, Julie, Sylvan and Dillon, Chantal and Michel, Lorenza and family, Aunt Jenny, and the Arévalo, Alden, Van Acker, Seck, Kirouac, and Lemay families.

To Carmen Elisabeth for the enduring friendship and amazing illustrations; Paul Healey and Hannah Brook Farm crew; New Moon Organics family; Carmen Rosen, Bea Edelstein, and Still Moon Arts Society; Forest Stearns; Jah Chupa and Dread Lion Radio; Costanza, Roberto, Benedetta, and the Italian Cultural Centre of Vancouver; Rick Havlak and Homestead Junction; Alicia Maraván and Guapamacátaro Art and Ecology Center; Nancy Cottingham Powell and North Van Arts; Lucas King and the Art and Fungi students at Mountainside Secondary School; my supportive editors Rob West, Linda Glass, and everyone at New Society Publishers; and the communities of Humboldt County, Vancouver, the East Bay, Lasqueti Island, and Powell River.

To the Coast Salish Peoples—the səlilwətaʔɬ (Tsleil-Waututh), Skwxwú7mesh (Squamish) and xʷməθkʷəy̓əm (Musqueam) Nations—on whose unceded traditional territories I live, and to the Wiyot and Yurok Peoples, on whose traditional territories I grew up, for their long history and continued stewardship of the soil, waters, air, and ecosystems.

INTRODUCTION

MUSHROOMS AND HUMANS: PAST, PRESENT, AND FUTURE

Mushrooms as we know them today evolved at least 120 million years ago—well before the time of the dinosaurs—and they have been part of our lives for as long as humans have existed. One of our closest living relatives, the mountain gorilla, is a passionate mushroom eater. The earliest record of humans eating mushrooms comes from a burial site in Europe from about 18,700 years ago, and the oldest record of mushroom cultivation by humans dates from about 600 CE in China. In Europe, button mushrooms (*Agaricus bisporus*) were first propagated in quarries and caves in France around 1650, on composted horse manure. The practice of fungiculture has developed rapidly since pure culture laboratory techniques were developed around the turn of the 20th century; these lab techniques require a high level of sanitary measures that are not easily achieved at home on a low budget. But things have changed, and it is now easier than ever to grow your own mushrooms. The proliferation of information-sharing via online forums has decomposed many of the barriers to DIY mushroom cultivation. Widely dispersed communities of home-scale cultivators have developed simple technologies (aka *teks*) to facilitate growing mushrooms at home.

◀ Gathering mushrooms is an age-old practice. JA SCHINDLER

Uma Echo Kirouac Arevalo enjoys the aroma of her harvest of pink oysters. WILLOUGHBY AREVALO

Certainly, people grow mushrooms for many different reasons. They are well loved as food, and most cultivated mushrooms are grown for this purpose. Practically all species of edible mushrooms have medicinal properties, as do many species that are inedible due to their texture or flavor. People are becoming increasingly aware of the health benefits that can be gained by consuming mushrooms regularly, including immune support, cancer prevention and treatment, and more. Incorporating mushroom cultivation with other forms of agriculture provides opportunities to build soil, cycle "waste" products, increase biodiversity, and boost ecological resiliency. And the application of cultivated mushroom mycelium into polluted land has been increasingly studied and implemented in *bioremediation*—the practice of engaging living agents to break down or remove toxins from soil, air, and water. Fungi produce unique enzymes capable of degrading some of the most toxic chemicals that humans have created into benign molecules. Fungi invented these enzymes about 299 million years ago to break down *lignin*, the rigid, durable component of wood that had recently (in an evolutionary time frame) been invented by plants, the buildup of which was choking out their ecosystems. Many DIY mycologists envision a near future in which fungi, with the help of humans, invent enzymes for decomposing the vast accumulation of plastics that we have created. And, of course, mushrooms are a pleasure to the senses. They offer us forms, textures, aromas, and flavors that remind us of other beauties while being uniquely their own.

A BIT ABOUT ME AND MY APPROACH TO MUSHROOM CULTIVATION

My love for fungi came long before I began growing them. I grew up in the humid coastal redwood ecosystem in Northern California, where mushrooms are abundant and diverse. I was fascinated by age four. I read as many mushroom books as I could get my hands on, and often brought specimens home to study. By age 13, my supportive—but not mycologically savvy—parents trusted my skills

enough to let me cook and eat the edible mushrooms I was able to confidently identify. At Humboldt State University, I was finally able to study mycology formally, though I majored in art, and it was then that foraged mushrooms became a significant part of my diet.

After reading *Mycelium Running* by Paul Stamets, I realized that my relationship with mushrooms was not very reciprocal. I studied mushrooms. I hunted mushrooms. I ate mushrooms. And in doing so, I objectified mushrooms. So, I made a conscious effort to give back to the fungi that I loved so much—by learning how to grow them and teaching my peers about them. I began by teaching a mycology workshop at a skillshare at a punk house in my town, and I soon began leading educational forays in the woods.

I approached Mike Egan, the mushroom grower at my farmer's market and asked him for a job doing inoculations. Three days later I arrived at Mycality Mushrooms, freshly showered and ready to work in the lab. Through my training with Mike, I learned the ins and outs of commercial gourmet mushroom production. Amazingly, despite my lack of cultivation experience, Mike trusted me to perform the most sensitive and technical part of the cultivation process. With my enthusiasm and his support, we expanded the number of species his farm was producing from four to over a dozen. While my main task at the farm was inoculating the fruiting substrates with grain spawn, I got the chance to try my hand at just about every step of the cultivation process.

During my two-year stint working at Mycality, I began touring the West Coast to teach workshops, and I had the opportunity to present at the 2012 Radical Mycology Convergence in Port Townsend, Washington. I had never experienced anything like this gathering of mushroom nerds, activists, cultivators, and visionaries. I found myself instantly enmeshed in a dynamic community of sharing and mutual support — reflective of the interspecies communities of symbiotic fungi. It was through this network that I began to learn about home-scale mushroom cultivation practices.

Less than a year later, I followed the love of my life to Vancouver, Canada, where we moved into a tiny cabin in a wild garden behind a chaotic but charming clown house. I found myself starting over in mushroom cultivation, with 100 square feet of shared living space, no mushroom cultures, no lab, no job, and very little equipment. I slowly adapted to my new constraints and built up a simple but effective cultivation setup that allows me to grow a modest amount of mushrooms for my family to enjoy.

I practice and teach contemporary, low-tech methods of mushroom cultivation. These were developed in part by the online mushroom community and introduced to me by *Radical Mycology* author Peter McCoy and others I met at the Convergence. These methods make it easier than ever for people to grow mushrooms at home with less dedicated space, less specialized equipment, and less infrastructure cost.

▲ Me at age 14 with my mom, Lauraine Alden, our friend Marco Gruber, and the day's pick of chanterelles. ELSA EVANS

My personal tendency is toward a fairly loose cultivation practice, similar to how I play music by ear, improvise, and cook mostly without adhering strictly to recipes or precise measurements. I implement minimal environmental controls and situate my mushroom gardens in the microclimates that already exist. I opt for scrappy salvaged materials over sleek purchased ones. As a hobbyist, educator, laborer, artist, and parent, I cultivate when I can, rather than on a strict schedule. My hope is that this book will provide you with the basic skills, information, and strategies needed to build your own cultivation practice—one suited to your own personality, living situation, and intentions. Through my ever-evolving relationship with fungi, both cultivated and wild, I have learned great lessons about reciprocity, and the fungi have also led me into many nurturing relationships with other *mycophilic* (mushroom-loving) humans. I see cultivated mushrooms as companions, friends, and members of my personal ecosystem—my interspecies community—and I of theirs. So, while the book is called *DIY [Do It Yourself] Mushroom Cultivation*, the truth is that we *Do It Together*.

MUSHROOM BASICS

▶ The features of a mushroom's anatomy. CARMEN ELISABETH

WHAT ARE MUSHROOMS?

Mushrooms are the fruiting bodies of *mycelium,* a network of threadlike cells that is the vegetative body of the fungus. I think of mushrooms as temples of sex: ornate and highly organized structures that emerge from (and of) mycelium to create and disperse spores—the analog of our eggs and sperm. Mushrooms arise only when conditions are conducive and when the mycelium recognizes sexual reproduction as a priority for the devotion of energy and resources. Once their spores have been dispersed, the fruiting body withers.

Mycelium can persist in its substrate as long as it has adequate resources and is not attacked, eaten, or otherwise destroyed. Depending on the circumstances and species, this can be as short as months or as long as millennia. Meanwhile, most types of mushrooms are evanescent—some emerge, sporulate, and decay within hours. More last a number of days to weeks, and some live for months, years, or even decades.

Taxonomy and Classification of Mushrooms

Fungi are a large and diverse group of organisms that are classified separately from plants, animals, bacteria, and protists, forming their own kingdom. Mushroom-forming fungi exist in two of the seven broad divisions (phyla) of fungi: the Ascomycetes and the Basidiomycetes.

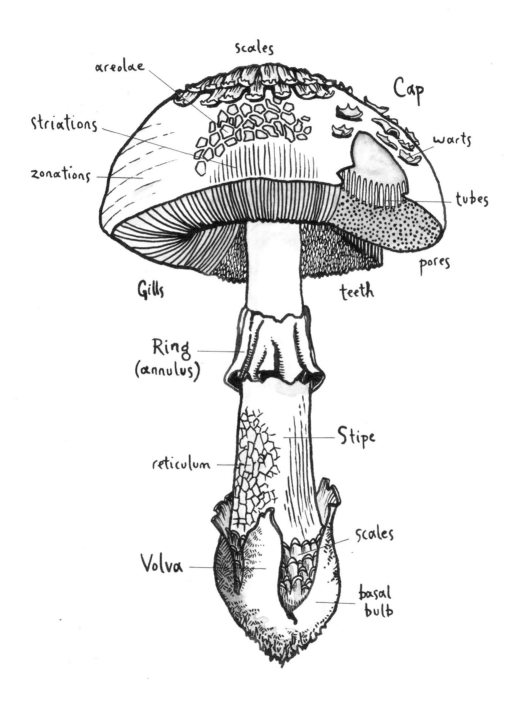

Lobster Mushroom

Caterpillar Fungus

Grey Morel

Black Truffle

▲ Ascomycetes CARMEN ELISABETH

Ascomycota, the sac fungi, is a very large and diverse group that includes some mushrooms; however, most Ascomycetes take forms other than mushrooms. The group includes many unicellular yeasts, including *Saccharomyces cerevisiae*, the species we have to thank for bread, beer, and wine. Many are molds such as *Penicillium*, *Aspergillus*, and *Fusarium*.

Common Name	Human	Caterpillar Fungus	Shiitake
Kingdom	Animalia	Fungi	Fungi
Phylum	Chordata	Ascomycota	Basidiomycota
Class	Mammalia	Sordariomycetes	Agaricomycetes
Order	Primates	Hypocreales	Agaricales
Family	Hominidae	Cordycipitaceae	Marasmiaceae
Genus	Homo	Cordyceps	Lentinula
Species	sapiens	militaris	edodes
Binomial	*Homo sapiens*	*Cordyceps militaris*	*Lentinula edodes*

While humans have succeeded in cultivating some of the edible and medicinal Ascomycete mushrooms such as morels (*Morchella* spp.), caterpillar fungus (*Cordyceps militaris*), and some truffles (*Tuber* spp.), most of these require advanced techniques, so they will not be the focus of this book. Although many of the techniques presented here can be applied to the cultivation of their mycelium, advanced techniques are required for fruiting them.

The phylum Basidiomycota, the club fungi, includes the great majority of mushrooms, such as all the mushrooms with gills, pores, or teeth, and most of the jelly and coral fungi. Nearly all the mushrooms that are grown by humans are in this group, so we will focus on their life cycle and biology.

A Generalized Basidiomycete Mushroom Life Cycle

Spores (center and 12 o'clock) are miniscule propagules. Like sperm and eggs, spores are *haploid*, containing only half a set of genes (and they lack an embryo, which seeds have). Unlike sperm

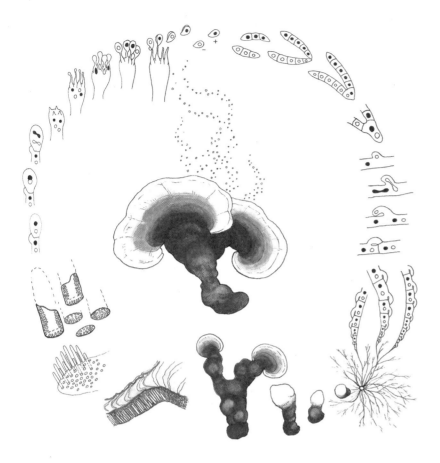

▲ The life cycle of reishi (*Ganoderma lucidum*). Bold terms follow the image, clockwise from top. CARMEN ELISABETH

and eggs, spores can begin to grow without being fertilized by one another.

Germination (12:30) occurs when a spore finds itself in conditions that will support its growth: i.e., the right water, food, air, and temperature. The spore then pushes out a threadlike cell called a *hypha* (plural: *hyphae*), which grows longer and branches, forming permeable cross-walls (septa) to control the flow of cell contents. A mass of hyphae is referred to as *mycelium*. As this mycelium grows, it exudes pheromones for communication with other fungi. These pheromones, similar to those of humans and other animals, call out

Fly Agaric

Turkey Tail

Lion's Mane

Arched Earthstar

▲ Basidiomycetes CARMEN ELISABETH

Mushroom Basics • 9

> Most cultivation books use the term *colonization* rather than *myceliation*, likely a relic of the coincidence of developments in biology with the expansion of European peoples across much of the world. But I don't want to valorize the history of colonization and the continuing subjugation and oppression of indigenous peoples and the land upon which they live, nor do I want to alienate folks whose cultures and land have been colonized. We must shift our way of speaking.

chemically to potential mates. When a compatible mate is sensed—the pheromone fitting lock-in-key into the receptor—attraction (1 o'clock) is established, and the two mycelia grow toward each other.

Fusion (2 o'clock) occurs when the two mycelia meet and become one mycelium. Their nuclei migrate (3 o'clock) into the mycelium of each other, until each cell contains one nucleus from each parent spore. This dikaryotic (two-nuclei) phase (3:30), which is fleeting in our species (the brief moment when the sperm and egg fuse), forms most of the mushroom life cycle. In cultivation, this genetically unique, mated mycelium is referred to as a *strain*.

Myceliation (vegetative growth, 4 o'clock) continues as the mycelium exudes enzymes to break down molecules in its *substrate* (the material it lives in and eats) and absorbs the components it needs to build more hyphae and grow. Its immune system produces extracellular antibiotics, antioxidants, and other chemicals to protect itself from viruses, bacteria, other fungi, *mycophagous* (fungus-eating) invertebrates, and oxidative stress.

Once sufficient energy is accumulated, and when environmental conditions are right, the mycelium is finally capable of producing mushrooms. They begin as dense hyphal knots (4:30), which then differentiate into primordia (5 o'clock), also called *pins* or *pinheads* by some cultivators. The mushrooms develop (6 o'clock) as the mycelium pumps the primordia full of nutrient-rich cell fluids, causing stalks to elongate and caps to expand (center).

As the fertile undersurface (gills, teeth, tubes, etc.) matures (6:30–8 o'clock), waves of the sexual cells called basidia (9 o'clock) develop on its surface. Within each *basidium,* the two nuclei fuse. Their genes combine, and they divide (10 o'clock) into four new nuclei (10:30), each with a unique half-set of genes. These nuclei migrate to the outside of the mature basidia (11 o'clock). Dosed with just enough nutrients to fuel the initial germination, these new *spores* are liberated by ballistospory (11:30). Once in the open air, they are carried on the breeze, hopefully landing on fertile substrate near compatible mates. Because relatively few spores will live to

▲ Which came first? The (chicken of the woods) mushroom or the spore?
CARMEN ELISABETH

complete their life cycle, each mushroom may produce millions to trillions of spores.

Queering the Fungal Queendom

I've just described how most mushrooms live and reproduce, but there are many exceptions. While most species' spores must breed with another spore (*heterothallic*) as shown in the illustration, about 10% of mushroom species' spores are self-fertile (*homothallic*). Often, more than one factor determines *mating type* (what we think of as "sex"), meaning that many mushroom species have more than two mating types. The extreme example is the split gill fungus *Schizophyllum commune*. It has 23,328 mating types, each compatible with 99.98% of the others. This has given it remarkable adaptability.

While aerial spore dispersal (ballistospory) is the norm, some mushrooms have evolved internal spore production and alternative spore dispersal strategies that involve animals, raindrops, or other agents. These are called the *gasteroid* fungi, because they make their spores inside stomach-like structures. Wait. Do you know of any other beings that make their reproductive units in stomach-like structures? I think we should call these species the *uteroid* fungi. Several uteroid mushrooms are cultivated, including some truffles and stinkhorns.

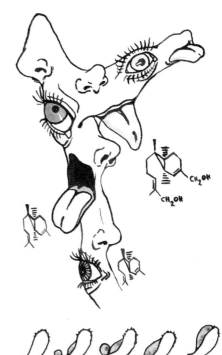

▲ On the surface of the hyphae are sensory receptor molecules—similar to those we use to see, smell, and taste—some of which receive pheromone signals from potential mates. CARMEN ELISABETH

▲ Two water droplets condense on each spore and grow until they converge. The sudden shift in balance hurls the spore into the air at up to 25,000 g's. This is the fastest acceleration known in nature. Being so tiny, air drag causes the spores to decelerate rapidly too, slowing to a freefall as soon as they're off the wall of basidia. CARMEN ELISABETH

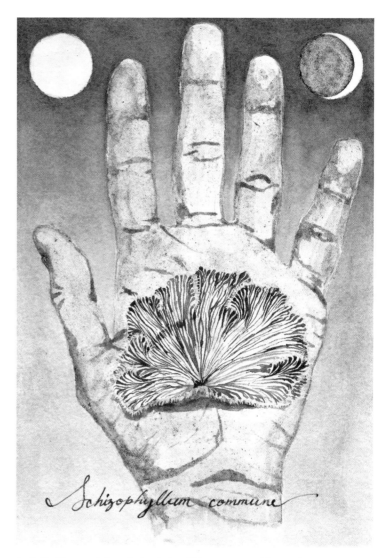

▲ Split gill's sexual diversity has helped it become one of the most common mushrooms in the world. It has been widely cultivated for genetic studies, though it is not commonly grown as food or medicine despite being edible and medicinal. WILLOUGHBY AREVALO

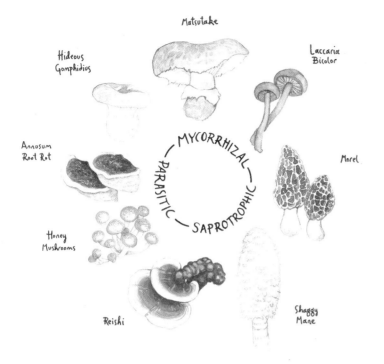

Clockwise from top: Matsutake (*Tricholoma magnivelare*) is mycorrhizal. *Laccaria bicolor* decomposes animal carcasses and hunts microscopic springtail animals, giving the nitrogen to their mycorrhizal tree partners. Morels (*Morchella* spp.) can be saprotrophic and/or mycorrhizal, and potentially weakly parasitic, switching over time. Shaggy manes (*Coprinus comatus*) are saprotrophic, but they also trap and eat nematodes. Reishi (*Ganoderma* spp.) are usually saprotrophic but will act as facultative (weak) parasites on an already compromised tree. Honey mushrooms (*Armillaria* spp.) are aggressive parasites but will act as saprotrophs after host death, or as needed to travel. Annosum root rot fungus (*Heterobasidion* spp.) is parasitic but keeps eating its host after death. Hideous gomphidius (*Gomphidius glutinosus*) is parasitic of mycorrhizal *Suillus* mushroom mycelium.
CARMEN ELISABETH

MUSHROOMS IN ECOSYSTEMS

Mushrooms and their mycelium play a variety of crucial roles in land-based ecology, with mushroom species evolved to fill niches in nearly every ecosystem on Earth. There are three main categories into which mushrooms are grouped based on their nutritional mode (energy source). However, many mushrooms can play more than one role, so I like to think of the nutritional mode as a spectrum on which there are three nodes: mycorrhizal, saprotrophic, and parasitic. Each of these is described below.

Many fungi have co-evolved with other organisms over millions of years to live in mutualist symbioses (*sym*=together/*biosis*=way of living) such as *mycorrhizae* and *endophytes* (fungi growing inside plants), and lichens. These symbiotic fungi get carbon (sugars) from their photosynthetic partners (plants, algae, and cyanobacteria), and most provide mineral nutrients, water, and/or protection in return. Many fungi are are symbiotic with bacteria; most of these

> The scientific names of mushrooms are constantly changing as molecular phylogenetic analysis reveals unexpected relationships. This causes a lot of confusion for mushroom people. I use the most current names at the time of writing, but when that's different from the best known name, I include the synonym in parentheses.

relationships are not well understood by science. When we cultivate mushrooms, especially at home, we enter a mutualist relationship with fungi, exchanging substrate, shelter, and more for food, medicine, and beautiful companions.

Mycorrhizal Fungi (*mykes*=fungus/*rhiza*=root)

Mycorrhizal fungi engulf and infiltrate plant root tips, exchanging essential minerals and water for photosynthesized carbohydrates. There are several types of mycorrhizae that do not form mushrooms and one that does, the *ectomycorrhizae*. Many attempts have been made to cultivate ectomycorrhizal mushrooms such as chanterelles, matsutakes, boletes, and truffles; results have been inconsistent, at best. While some of the techniques offered in this book may apply to the culturing and spawn production of certain mycorrhizal species, the topics of establishing a symbiosis and fruiting mycorrhizal mushrooms are beyond the scope of this book. The arbuscular mycorrhizal fungi (AMF, or *endomycorrhizae*) do not form mushrooms but are known for their generous support of most species of land plants, including the majority of garden and farm crops. While AMF can be grown fairly easily on a small scale and low budget, this, too, is beyond the scope of this book.

▲ Honey mushrooms fruit prolifically from trees they have killed. "Let us therefore trust the eternal Spirit which destroys and annihilates only because it is the unfathomable and eternal source of all life. The passion for destruction is a creative passion, too!"

MIKHAIL BAKUNIN Russian political philosopher, anarchist, and atheist, 1842 WILLOUGHBY AREVALO

Parasitic Fungi: Those Who Eat at Another's Table

Rather than inhabiting plant tissues and giving back like the mycorrhizals do, the *parasites* take energy from their host plant without any direct reciprocation. However harsh as this may seem, parasitic fungi can enhance their ecosystems. They help to drive succession by killing weak trees, which opens up the forest canopy, creates new habitat niches for amimals, fungi, and plants, and provides selective pressure toward the evolution of more resilient tree species. Some parasites kill their hosts and continue to eat them after they die, while others depend on their host's survival and do not kill. Parasitic mushrooms are not often cultivated, as many people are wary of their power and potential for destruction, but some are cultivated and applied as biological controls for invasive plants.

Saprotrophic Fungi (*sapro=rot/troph=eating*)

Saprotrophic fungi are Earth's great decomposers, working together with invertebrates and bacteria to break down the world's dead plant matter and convert it into smaller molecules, carbon dioxide, and water that can be recomposed into other forms of life. In so doing, they connect the cycle of life and death. They play a huge role in building soil, breaking hard-to-digest woody biomass into bite-sized pieces for invertebrates and bacteria to break down further. Because of the simplicity of preparing and feeding dead plant matter (called *substrate*) to the mycelium, these mushrooms comprise the vast majority of mushroom species cultivated, and they will be the focus of this book.

Saprotrophic fungi thrive in a wide spectrum of environmental niches, ranging from wood, to leaves, to humus. Some species are very particular in their substrate needs, while others are generalists and can eat a wide range of substrates. For example, wild brick tops (*Hypholoma sublateritium*) grow almost exclusively on oak and chestnut stumps, while oysters (*Pleurotus* species) grow naturally on a wide variety of woods and have been successfully cultivated on over 200 types of agricultural and forestry wastes.

Within a particular piece of wood, there is a succession of different decomposers: *primary decomposers* are parasitic or aggressive saprotrophic mushroom species that eat their fill before leaving it to the *secondary decomposers,* which eat partially degraded wood. Eventually, the rotten wood is transformed to humus, which is inhabited and eaten by the *tertiary decomposers*. At each step in the process, the components in the wood are broken into smaller and smaller molecules until they are ready to be taken back up into plants, usually with the help of mycorrhizal fungi. Most of the wood-rotters that are commonly cultivated are primary decomposers,

▲ Spontaneous fruiting of *Lenzites cf. elegans* from a crude-oil-soaked log in the Ecuadorean Amazon rainforest. Amisacho and CoRenewal are two non-profits collaborating with each other and with fungi, bacteria, plants, and humans to bioremediate the numerous oil pits left throughout the region by Chevron/Texaco.
MAYA ELSON

preferring fresh wood that has not been inhabited and eaten by other fungi. These fungi are the focus of this book. The cultivated species that are naturally found on the ground growing on non-woody substrates such as leaf litter, compost, manure, or humus are mostly classified as tertiary decomposers. For example, blewits (*Clitocybe nuda*) are tertiary decomposers that eat leaf litter in the wild but can be grown on straw that has already been partially eaten by oysters.

SAPROTROPHIC MUSHROOM NICHES

Wood-Rotting Fungi

The wood-rotting fungi, whether saprotrophic or parasitic, are also classified by the type of rot that they produce. Wood is made up of a matrix of stringy, fibrous cellulose and hemicellulose running longitudinally through rigid, brittle, and porous lignin.

The *brown-rotters* are able to work around the hard-to-degrade lignin and eat away the cellulose and hemicellulose, leaving behind a blocky, crumbly brown rot. These porous blocks of residual lignin can remain in soil for many years, increasing the soil's water and air retention as well as its *cation exchange capacity*—the ability for the upper layers of the soil to hang onto nutrients rather than letting them be washed away into the groundwater—allowing plants greater access to these nutrients.

Relatively few brown-rotters are cultivated. These include chicken of the woods (*Laetiporus* spp.), birch polypore (Femitopsis betulina [*Piptoporus betulinus*]), shimeji, and elm oyster (*Hypsizygus* spp.).

Most cultivated mushrooms are *white-rotters*. They give the wood a pulpy, bleached-out appearance by degrading the lignin in order to access the cellulose and hemicellulose within the matrix. The powerful enzymes produced by the white-rotters to degrade lignin have the ability to degrade petroleum hydrocarbons and other toxic chemicals.

Litter and Compost Decomposers

Some fungi eat litter (in the sense of leaves, twigs, humus, and other smaller plant particles), and some of these fungi may also be able to eat human-made litter, such as paper products and certain agricultural wastes, including used substrates from the cultivation of other mushrooms. Some mushrooms are very substrate specific, such as those small mushrooms that grow exclusively on the cones of a certain tree species. These mushrooms are also feeding on lignocellulose, though in an easier form to digest than wood. They often rely on a microbially rich ecosystem in which to thrive, and some are weak or slow-growing in pure culture. Therefore, most of these grow better in gardens or other more natural settings, rather than in containers indoors. Some of the cultivated members of the group are button mushrooms and other *Agaricus* species, shaggy manes (*Coprinus comatus*), parasols (*Macrolepiota procera* and *Chlorophyllum* spp.), blewits (*Clitocybe nuda,*) and paddy straw mushroom (*Volvariella volvacea*).

WHAT MUSHROOMS NEED TO SURVIVE AND THRIVE

Air

Just like us, fungi consume oxygen (O_2) and produce carbon dioxide (CO_2) through their metabolism. Lacking lungs, they respire with all their mycelium on a cellular level. They are more sensitive than we are to relative levels of $O_2:CO_2$. Mycelial growth doesn't require as much oxygen as fruiting does, though without any access to oxygen, mycelium will cease to grow, suffocate, and die. Imagine mycelium growing through the interior of a log. There is little airspace and airflow from within the wood to the outside air, so CO_2 levels can get pretty high. When the mycelium is preparing to fruit, it grows to where there is more oxygen, at the surface of the wood, perhaps in a fissure in the bark. It senses the oxygen in the ambient air, which helps to trigger the formation of fruiting bodies. In our substrates, particle size is important; if the particles are too large, there will be big air gaps that the mycelium will have to bridge, which is energetically expensive. If the particles are too small, over-compaction and restriction of airflow causes anaerobic conditions.

Water

Water is life, and mycelium and mushrooms are mostly water, roughly 90%. Because of this and because of how they exude chemicals (sometimes called *secondary metabolites*) to surround themselves as they grow, water can be a major limiting factor for mycelial growth. Water must not only be present in a substrate, but it must also be available. Each particle of a substrate should be thoroughly moist and coated in a thin film of free water, so it can be easily utilized by the mycelium. However, if the substrate is so wet that water pools in the bottom of the vessel, then airflow will be compromised and mycelial growth along with it, favoring instead bacteria and molds. As you get to know your substrate materials, you will develop a sense for how moist they should feel.

Water quality is also an issue. Most cultivators use tap water, which in most areas is okay. The levels of chlorine (or chloramine, a non-evaporating chemical that is increasingly substituted for chlorine) in most tap water are not high enough to do major damage to our fungi, but if you are able to dechlorinate your water by simply letting it sit out uncovered overnight, that is a good practice. Chloramine can be removed from water by various chemical methods. Mineral-rich water such as spring water is great, if you have access to a source, but I do not recommend buying spring water or other bottled water for growing mushrooms unless your tap water is significantly polluted. Because many mushrooms hyperaccumulate heavy metals, water contaminated with these elements should not be used. If you filter your water for drinking, I would advise filtering your mushrooms' water as well.

Food

Like animals, fungi are *heterotrophic*—they get their energy (carbohydrates) from outside sources. Many have evolved to produce strong enzymes that can break down and get energy from tough stuff like the lignocellulose matrix of wood, and many can break down rocks and absorb the minerals. Think of the human digestive system as an outside-in variation on a theme that evolved in our common ancestors. We ingest and digest; fungi *"outgest"* and digest.

Decomposer mushrooms can be grown on a wide variety of plant-based substrates. Some species are restricted to specific types of woods, while others can eat and grow from many types of lignocellulose-rich plant matter. All dry land-plant matter contains lignocellulose.

In addition to carbohydrates, fungi need nitrogen from and for proteins, and small amounts of micronutrients such as calcium, sulfur, lipids, and vitamins. Most of the commonly used substrates supply ample amounts of these nutrients to support fruitings, but not all substrate ingredients are created equal. You could grow oyster mushrooms on logs, sawdust, or paper —all of which are wood products—but the nutrient density and quality of the substrate (and the mushrooms produced on them) decrease the farther you get from the source. Keep these nutritional needs in mind when experimenting with substrate formulation, and when in doubt, mimic the natural substrate of the mushroom.

Warmth

Mushrooms are *mesophiles,* like us; they like middle-range temperatures—few thrive in freezing cold or sweltering heat. Mycelial growth halts when it gets too cold, and mycelium can die of overheating. However, even in extreme climates, human companions of fungi can create microclimates in which mushrooms can be happy, and our living spaces are full of these niches. There is a fairly wide range of preferred temperatures for growth and fruiting for various species—between 45–100°F (7–38°C), which is about the temperature range inside a home. Each species (and even each strain) of mushroom has its own temperature preferences, so with a good collection of cultures and some forethought, one can fruit mushrooms indoors and/or outdoors year-round in nearly any climate.

Fruiting Surface

Different species have different needs for their spatial relationship to gravity and the surface from which they grow. Their shape and natural growth habit reveal these needs. Ground-dwelling mushrooms normally fruit from the top of a horizontal surface. Many wood-rotters prefer to fruit from the side of a vertical surface, but some will fruit from side, top, and even bottom surfaces of their substrate. For species like shiitake, a broad fruiting surface should be exposed to the air for mushrooms to fruit all over. For others (like oysters), small holes in the vessel allow focused fruiting in clusters, which prevents moisture loss in the substrate and facilitates harvest.

OVERVIEW OF THE CULTIVATION PROCESS

CULTIVATION FLOW

Mushroom cultivation is a series of exponential expansions of mycelial mass, culminating in a shift in environmental conditions that encourage the formation of mushrooms. Each cultivator has a slightly different process, tailored to their situation. Many growers skip the culture work and spawn production, using purchased spawn instead to inoculate beds, logs, and other fruiting substrates. The process generally goes a little something like this:

Step 1: Culture Creation

Every hyphal thread is a stem cell capable of growing into a whole organism, including those in a mushroom. A small sample of tissue is taken from the inside of a mushroom and transferred into a sterilized *growing medium*. Traditionally, this is nutrified agar medium in a Petri dish, but I use liquid culture medium in a jar. The sample is *incubated* (kept warm) while the mycelium grows over or through the medium; this step lasts for several days to weeks. Alternately, spores are germinated in a sterilized growing medium, and a *strain* (mated pair) is isolated from the many that develop.

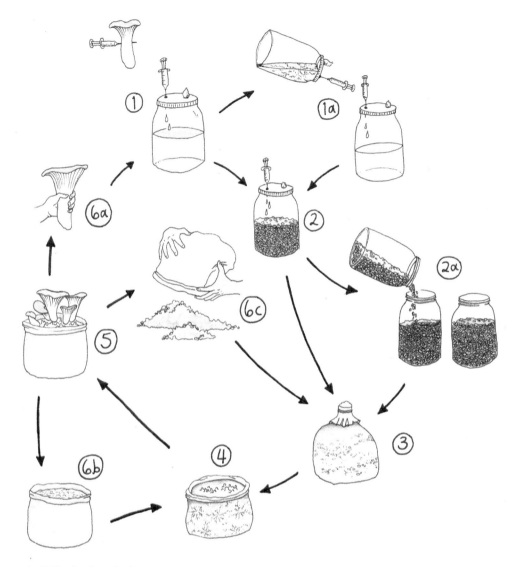

▲ Cultivation flow chart. CARMEN ELISABETH

Step 1a: Culture Expansion (Optional)

When the medium has been mostly but not completely myceliated, a little bit of culture is transferred to additional containers of sterilized culture media. Because mycelium grows from the tips, samples are taken from the leading edge, where the life force is strongest and most active. Like the tissue culture, these subsequent cultures are incubated while they grow.

Overview of the Cultivation Process • 21

Throughout the process, all vessels should be labeled immediately after inoculation. Labels may include inoculation date, species (usually abbreviated, for example: PO for *Pleurotus ostreatus*), strain, generation, inoculum source, date of shaking or other action, substrate formula, and any experimental variables. Each cultivator develops their own shorthand code to minimize the amount of time spent labeling.

It is helpful to keep a cultivation journal to track your cultivation practice. You may want to record things like inoculation dates and rates, substrate formulas and preparations, occurrence of contaminants or pests, dates of initiation, dates and weights of harvests, environmental conditions, experimental variables, and other factors.

Step 2: Spawn Generation

Mushroom starter culture (either liquid or agar) is transferred into vessels of cooked and sterilized grain. The grain is incubated for a few weeks until fully myceliated, becoming *grain spawn*. Spawn is a carrier that moves the culture into the fruiting substrate. Grain provides the mycelium with an excellent nutrient boost to carry it into the next phase, like a big hearty breakfast.

Step 2b: Spawn Expansion (Optional)

The cultivator may expand spawn to more grain or to sawdust and incubate until myceliated.

Step 3: Spawn Run to Fruiting Substrate

The spawn is distributed amongst a number of containers containing a fruiting substrate such as sawdust, straw, compost, or other waste products, like coffee grounds. This is incubated while the mycelium from the spawn runs through the substrate. This takes one week to several months, depending on species and temperature.

Step 4: Initiation (Primordia Formation)

The containers of fully myceliated fruiting substrate are moved to the fruiting space, where the ambient temperature is lowered, the humidity is increased, the carbon dioxide level is lowered by increasing fresh air exchange, and light is introduced or increased. These environmental cues, along with the declining availability of nutrients in the substrate, trigger the formation of primordia (baby mushrooms, aka pins). This important phase is transitional, usually lasting several days to a couple weeks.

Step 5: Fruiting

Fresh air and gentle light are maintained. Temperatures can be a little warmer and humidity may be dropped a little compared to the initiation phase, but conditions should remain cool and quite moist. Mushroom development takes days to weeks.

Step 6a: Harvest!

Mushrooms are ready to pick when they're mostly expanded but the margins of the caps (if present) are still curved under. Hands are used to twist mushrooms free from the substrate. Clustered mushrooms such as oysters should be picked as whole clusters.

Step 6b: Resting

The mycelium will need time to recover after the great energy expense of mushroom production. Over a couple of weeks to months, the mycelium will further digest the substrate, rebuilding its energy stores. Reducing watering during the rest phase may be beneficial.

Initiation, Fruiting, and Harvest Repeated

Ideally, each container or block should be able to produce two or more flushes. In some cases, this is as simple as keeping it in the fruiting space and waiting for the mycelium to self-initiate. In other cases, blocks or logs should be soaked and/or cold shocked to be re-initiated. Each harvest is smaller than the last, but sometimes mushrooms appear over and over and over. I have watched flushes of oysters emerge from big bags of straw every few weeks for over half a year, and logs can produce for years.

Step 6c: Cycling of Spent Substrate

Between diminishing returns and the inevitable invasion of old fruiting blocks by molds and gnats, the cultivator must eventually thank the mycelium for its hard work and remove it from the fruiting space. A block may be re-sterilized or re-pasteurized and used to grow other mushroom species, or it can simply be left outside (where it may just give a big surprise flush). It can also be composted, fed to livestock, dried and used as animal bedding, or used to filter polluted water.

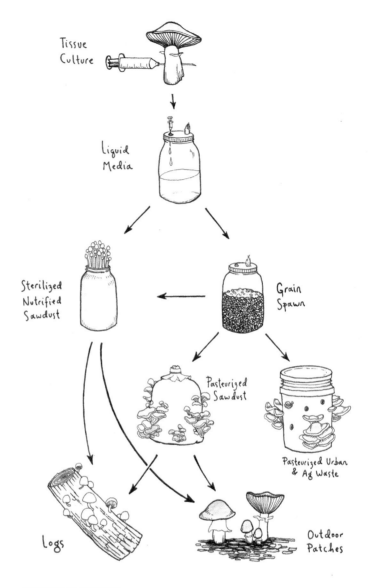

▶ The Mycelial Pathway. This is the mycelial pathway I recommend. The beauty is that it doesn't require an aseptic transfer space—except for the optional step of going from grain spawn to sterilized nutrified sawdust. CARMEN ELISABETH

PREPARATION

What Will You Do and How Will You Do It?

There are many ways to grow mushrooms, and how you do it will depend on your ambitions, resources, and capacity. Of course, you can just dive in and see what happens, but it can be very helpful to write down or at least think through a plan. Dreams are spores for growth, and I encourage you to dream big—but you'd be wise to start small while you learn what you are doing. Then your setup can grow organically in proportion to your skills, time, and space. Think about your goals, resources, and capacities, and build your skills and infrastructure so you can achieve your goals. What types of mushrooms do you want to grow and why? If your dream is to grow morels, which are notoriously difficult, you may want to start with wine caps instead, which also grow in the ground but are much easier to cultivate. If you want to bioremediate a giant brownfield, start on a small sample of the contaminated soil. Refer to Appendix 1 for good species to start out with. The more you study and learn, the more grounded you will be in your aspirations, which will make you more likely to succeed. A major factor in your productivity will be the amount of time you devote to your efforts. While fairly easy, mushroom cultivation is a labor-intensive practice.

Business Planning

If you want to make an income growing mushrooms, start by growing enough to share some with friends, then try to make new friends by sharing your harvest with potential customers. Explore your potential clientele. Try to establish a regular rhythm of production and harvest so you will be able to supply customers when *they* need mushrooms. Most sales outlets like farmer's markets, restaurants, and grocery stores will need a consistent supply once or twice a week. Nurture your relationships, and word will spread. Before long, you may need to boost production to fulfill the demand. For a mushroom farm to make money, all the elements need to be efficient and scaled to each other. Specialization and cooperation can increase profits—for example, it may be more profitable to buy spawn instead of making it yourself; or to make it for others and not focus on mushroom production. Most mushroom farms start out as cottage industry, and many of them stay that way. Indeed, many of the photos and concepts in this book were generously shared by home-scale farmers.

WORKSPACES, TOOLS, AND EQUIPMENT

Living with Fungi: Living Room, Bedroom, Bathroom, Mushroom

Space is a determining factor in production. As a home grower, one must think ecologically to find creative ways to use and share space with people, furniture, stuff, pets, houseplants, etc., situating the working, storage, and growing spaces into niches that exist within our living spaces. Those with a big house, a basement, a garage, outbuildings, or acreage will have an easier time fitting mushroom spaces into their world, but even the apartment-dweller can eke out space for their fungal companions. If you get to the point where you are setting up a farm with dedicated spaces, I suggest consulting Stamets and Chilton 1983, Stamets 1993, and Cotter 2014 (see Bibliography).

LAB INFRASTRUCTURE AND ASEPTIC TRANSFER SPACES: FLOW HOODS, STILL AIR BOXES, AND MORE

The lab is the cleanest place in the mushroom-growing operation. The heart of the lab is the *aseptic transfer space,* where the air is clean. Airborne contamination can be significantly reduced by using settled air (as in a still air box), HEPA filtered air (as in a flow hood), or a convection current (as in the *oven tek* or with a Bunsen burner).

▲ The lab space occupies the interface between living room and kitchen at Wildwood Ecology Labs in Powell River, BC. THEO ROSENFELD

The point of the clean lab is to allow the cultivator to open containers of sterilized culture media or substrate for inoculation with greatly reduced risk of introducing airborne contaminants. This is essential for agar work, grain-to-grain transfers, and inoculating nutrified sawdust with grain spawn.

While one can avoid the need for a clean air space by using liquid culture media for inoculating grain and nutrified sawdust, there are instances when a clean air space can be of great benefit. For example, some mushrooms are easily cloned to liquid culture with a hypodermic needle, but others are not and are best tissue-cultured to agar with a scalpel or tweezers. Many methods of strain development or assisted adaptation rely on the agar format as well, and spores are

▲ My friend Max's wooden glove box. It is well built, though I find these big gloves cumbersome on my small hands. WILLOUGHBY AREVALO

much easier to work with on agar than in liquid culture. These techniques are covered in Chapter 5.

Still Air Box: Settling the Air

A still air box (SAB) or a glove box (GB) are effective, cheap, and easy to build, and they are the best option for beginners. The principle is that with still air, contaminants will not fall into the work—if the user practices good technique (which includes not moving hands over the work). An SAB is just a clear box with armholes, while a GB has gloves attached, sealing off the space. I believe the best designs are simple, made from large, clear-plastic storage bins. Some heavy-duty boxes are made of sealed plywood or metal with plexiglass for the top, but these are not necessarily better. An SAB is arguably better than a glove box; in addition to allowing the user to flame-sterilize tools outside the box, it's simpler to build and more ergonomic. And attached gloves create a piston effect when moving around, sucking in dirty air. An SAB user can still wear gloves and/or Tyvek sleeves to contain microbes from the skin, or simply wash well and sanitize skin with alcohol. Gloves should be nonporous, well-fitted, and thin enough to maintain a good level of dexterity.

SAB Construction

You'll need a big, tall, see-through plastic storage box (at least 60 qt, but 100 qt capacity is good), a coffee can, stove, hot mitt, sandpaper, and clear packing tape.

1. Remove lid from box. With your box upside-down on your work surface, reach forward to the box with bent elbows as if you were working in it with hands shoulder-width apart. Make marks where your middle fingers touch the box—these will be the center of your holes. Trace the coffee can with a marker to mark the holes.
2. Moderately heat up the open end of the coffee can on the stove, and—working outdoors—melt holes in the box.
3. Use fine-grit sandpaper to clean up the edges of the holes.

4. If there are any holes near the handles, cover them up with packing tape.
5. Keep the lid—the box can be used to store supplies, or even incubating cultures and spawn.

To use an SAB or GB, clean and sanitize the inside of the box and the work surface, then place box on work surface—some cultivators lay it down on a clean towel. Spray a fine mist of lightly soapy water inside to precipitate out and immobilize any airborne contaminants. With alcohol and paper towels, sanitize all materials, including the outside of spawn and substrate containers and tools needed for the work session, then load the supplies into the box. (SAB users can keep some equipment like an alcohol flame beside the box.) Allow the air to settle for a few minutes before beginning work. Keep your movements slow to minimize air currents, yet work efficiently and deliberately. Never use an open flame and alcohol together in a glove box. Fireballs aren't healthy for mycelium or cultivators!

> **LAMINAR FLOW HOODS: FILTERING THE AIR**
>
> In a laminar flow hood, a fan forces air through a HEPA (high-efficiency particulate air) filter and across the workspace, pushing away airborne contaminants from the room. A flow hood is ideal for eliminating airborne contaminants because it's effective and easy to use. However, it is also relatively expensive, difficult to build, and bulky. Used flow hoods can be purchased online; however, prices range from hundreds to tens of thousands of dollars. Even if you make your own flow hood, it would still cost a few hundred bucks due to the cost of a high-quality HEPA. Plans can be found online.
>
> When working in a laminar flow hood, run the fan for an hour or more before starting. This will clear out any contaminants that have settled on the outside of the filter. Clean the workspace with a disinfectant before bringing in tools and materials. Always keep in mind the direction of clean airflow with regard to your movements and the placement of objects in the workspace. Avoid placing your hands or any objects upwind of open vessels, as contaminants may be blown onto the substrate. Avoid coughing, sneezing, blowing, or speaking over the work, as the clean airflow may not prevent these forceful movements of air from carrying contaminants into the work.

▲ Agar jars cool in a small flow hood equipped with a 12" × 12" HEPA filter, a 6" can fan, and a clear tote cut to fit as hood. This hood was made by workshop participants at Fungi for the People in Eugene, Oregon.
JA SCHINDLER

Ultraviolet Light: Irradiating the Air and Surfaces

Some people mount UV lights inside SABs or flow hoods to kill contaminants in air and on surfaces. Beware that exposure to UV can cause skin cancer and eye damage. If used, UV lights should be run and then turned off before starting work. Note that UV lights do not kill anything that is in the shadow of something else.

Convection Currents: Lifting the Air

Bunsen burners (or propane torches) are used in microbiology labs to create a small semi-sterile field where agar transfers can be performed. The heat of the flame and its consumption of oxygen cause an updraft that pulls burned air up over the workspace. Being small and imperfect, it is not often used for mycology. However, Pablo Chencho, a DIY cultivator from Mexico, told me that some growers there, lacking access to flow hoods, use a crescent-shaped lineup of multiple burners to create a larger semi-sterile field.

The online mushroom community riffed on the convection current, creating the *oven tek*, where transfers are done in a 150°F oven. I tried it once, and it was awkward and only partially successful. An SAB is a much better low-cost option in most situations. If you use an oven, remove its door for less awful ergonomics.

THE LAB ENVIRONMENT AND TOOLS

If a dedicated space is available, great, but many home growers do not have this luxury. Nearly anywhere can work as a lab space when inoculating spawn only with liquid culture and airport lids, or when using a still air box (SAB) for transfers, though a clean, ergonomic, and well-lit surface to work on is important. Despite kitchens having the highest contaminant spore load of any part of the house due to the presence of food, a clean kitchen can make sense as a lab space for many home cultivators with small spaces, especially when using a SAB or airport lids to manage the air vector. Kitchen counters are typically smooth and non-porous, hence easily cleaned. The stove, sink, pots, and utensils are used in preparing media and substrates

▲ The tools of the trade. CARMEN ELISABETH

and washing hands, tools, and vessels. The fridge is useful for storing cultures and mature spawn, though they can become contaminated from being stored with food and its residue, so they should be sealed within a clean plastic box or zipper bag before being put into a home fridge. If you have room, a small culture fridge is recommended.

For those setting up a small home laboratory with a flow hood, I present the following guidelines: The lab is the primary workspace in the process of cultivation; this is where the most sensitive actions happen, and it should be kept as clean as possible. An extra room, metal shed, walk-in closet, cellar, or pantry can become a functional home lab. The lab should be disinfected regularly, so non-porous walls that can be wiped down with 10% bleach solution are ideal. If you cannot avoid biodegradable building materials, coat them well with high-gloss paint. (But don't use paints containing antifungal compounds—they are toxic to both humans and mushrooms!) Floors gather most of the contaminant spores and bacteria through settling, so they should be kept clean and not be carpeted. The lab should be sealed off from the outside by a tight door (weather stripping can be added) or a zippered plastic lining. Ideally, the flow hood's air intake source is outside of the lab to create positive pressure in the room, pushing clean air out of any gaps and not letting dirty air in. The workspace should be ergonomic and well-lit to keep the body and eyes happy and to improve the ease of work.

▲ I keep a small tackle box stocked with all the small tools and supplies needed for liquid culture work. This way I can easily create, use, and share cultures on the go. I keep syringes in a plastic bag—sterile ones wrapped in foil, and dirty ones not. WILLOUGHBY AREVALO

A WELL-EQUIPPED LAB MAY HAVE MANY OR ALL OF THE FOLLOWING SUPPLIES

Infrastructure and Equipment:
- Laminar flow hood and/or still air box
- Pressure canner (for sterilizing media, substrates, and tools)
- Vacuum cleaner, ideally with remote output

Tools:
- Scalpels and spare blades (for tissue culturing to agar and transferring agar. I prefer #11 blades.)
- Precision tweezers (for tissue culturing tough mushrooms to agar or holding tricky mushrooms while taking a needle biopsy)
- Razor blades and/or sharp knife
- Inoculation loop (for inoculating agar with spores)
- Measuring spoons and flasks (metric)
- Digital gram scale (for measuring media ingredients)
- Magnetic stir plate and stir bars for agitating liquid cultures (optional)
- Impulse sealer (if using polypropylene bags. Some home vacuum sealers can suffice.)
- Alcohol lamp filled with denatured alcohol (or 99% isopropyl), propane torch, or Bacti-Cinerator (for sterilizing tools)
- 16-gauge luer lock needles (1½" or longer, 18 gauge can suffice)
- 10 ml and 60 ml luer lock syringes

Supplies:
- Paper towels
- Hand sanitizer (avoid triclosan-based)
- Aluminum foil (can be reused a few times)
- Polyfill (polyester fiberfill/pillow stuffing—for air filter material)
- Micropore medical tape (air filter material)
- Tyvek (air filter material)
- Permanent markers

- Lighters
- Cotton balls (small for cleaning injection ports, jumbo for air filters)
- Alcohol prep pads (for surface-sterilizing mushrooms to culture)
- 70% isopropyl alcohol in a spray bottle
- 1:1 and 1:10 bleach:water solution in a spray bottle (optional)
- Distilled water
- 3% hydrogen peroxide (food-grade is ideal, available at quality pharmacies in 29% concentration and diluted at home in distilled water)
- Disposable nitrile and/or reusable rubber dish gloves (optional)
- Luer lock syringe caps (great for sharing/transporting LC syringes)
- Marbles, broken glass pieces, crystals and/or Teflon-coated stir bars (as LC agitators)
- Liquid culture and/or agar media ingredients
- Gypsum
- Mason jars and rings
- Airport lids
- Petri dishes, 125 ml Mason jars or 200 ml flask-shaped liquor bottles (for agar)
- Parafilm, plastic wrap or tape for wrapping agar plates (if using)
- Test tubes for making culture slants (optional; slants are discussed in Chapter 5.)
- Polypropylene (PP) filter patch bags (for sterilized substrates, optional)
- Polyethylene bags (PE) (for pasteurized substrates)
- Clear adhesive tape (for repairing punctured bags)
- Toilet paper tubes cut into thirds (for filter and collar assembly)
- Rubber bands (for filter and collar assembly)
- Stiff wire (for sealing filter patch bags instead of impulse sealer)
- Lab notebook

GATHERING SPECIALIZED EQUIPMENT, TOOLS, SUBSTRATES, AND SUPPLIES

Some of the stuff cultivators use is common and easy to get. Other items are more specialized and take some looking for. I've shopped at pharmacies, scrapyards, hardware, garden, feed, kitchen, grocery, thrift, and big box stores. I've searched flea markets and the internet, and I dumpster dive behind ice cream shops, pizzerias, and fishmongers. You can spend a lot or not much, depending on how resourceful and patient you are. I have managed to get set up pretty well without buying much new stuff, but I've been lucky and patient. One of the first things you'll need is a pressure canner, available at department or hardware stores, or found used on websites like Craigslist. eBay has the best value on syringes, but needles are usually found only at veterinary pharmacies (which also have syringes), as their sale is restricted in many countries due to drug prohibition. Call around. There are websites dedicated to the sale of used lab equipment. Numerous mycology-oriented businesses sell cultivation supplies and tools online, but they may not have the best prices. For more info see Appendix 2: Resources.

WHERE TO INOCULATE BULK SUBSTRATES

Inoculation of larger-than-small amounts of bulk substrates such as straw, multiple pasteurized sawdust bags, coffee grounds, logs, and compost can be too messy, bulky, or cumbersome to do within a glove box or in front of a small flow hood, so bulk inoculations are best performed out of the lab. Without supplementation, the substrates just mentioned are hard enough to decompose that most competitors introduced by the air vector will usually not beat the spawn in the race to claim the substrate. And because fully myceliated grain spawn is coated with hyphae and antimicrobial compounds, it resists competitors and can be used in the open air. Nevertheless, contamination can and does happen with these substrates, so care must be taken to minimize air exposure time and contamination by other vectors. Try to situate your bulk inoculation place in an area where the floor is easily cleaned. Use a table with a smooth, non-porous surface that is easily cleaned and disinfected. Plastic sheeting can serve as a clean surface on a wooden table and can be reused indefinitely if cleaned and sanitized before and after use and allowed to dry before storage. Many cultivators, myself included, do at least some bulk inoculations outdoors. If working outdoors, avoid windy, dusty, or rainy conditions, and keep some distance away from substrate prep and composting areas. Ambient spore loads are lower at night, favoring night-owl cultivators.

WHERE TO INCUBATE GROWING MYCELIUM

The incubation space should be consistently warm (70–85°F [21–29°C]), moderately humid, fairly clean, and moderately lit (no direct sunlight) or dark. It need not have much airflow in or out, as mycelium needs only a moderate amount of oxygen for growth. Think of it as a nest—a cozy, safe, peaceful place where mycelium can grow undisturbed. Ideally, it is situated between the inoculation space and the fruiting space for easy movement and no cross contamination. It should be furnished with racks or shelves for maximizing space for housing the growing mycelium. As various sizes and shapes of jars, Petri dishes, bags, buckets, and trays may be incubated in the

◀ I incubate in the water heater and furnace closet, which ranges between 68–90°F (20–32°C). Not perfect, but good enough. WILLOUGHBY AREVALO

chamber, plan for this when building it out. Growing mycelium generates heat. Pasteurized substrates can be incubated in a tight space, even touching each other, but bags of grain or nutrified sawdust should always have air space around them to avoid overheating of the substrate. For existing in-home spaces, the top of the fridge, the water heater closet, the cabinet above the stove, the laundry room, a closet, or a high shelf can work.

To make a small incubator, get two plastic bins of the same size. In the first, place a 100-watt aquarium heater and fill partway with water. Put the second bin inside and push down until water spills out and the top bin is tightly settled (do this in a bathtub or outside). Hold it in place with zip ties or a strap. Place vessels to be incubated in the top bin, along with a thermometer, and lid it. Be sure to dial in the temperature on the heater before inserting the top bin.

When space is limited, one may incubate liquid cultures and grain spawn in the same space as fruiting substrates (I currently do and it works out); but, ideally, cultures (especially agar) and grain spawn are incubated in a cleaner space (perhaps the lab), separate from the dirtier fruiting substrates. Alternately, fruiting substrates can be incubated in the fruiting space, though this is less than ideal, as fruiting spaces tend to contain high numbers of contaminant spores and pests that may infect the vessels of growing mycelium. Also, the environmental conditions would need to change from incubation to fruiting. Whatever the space, it is important to monitor the mycelium to keep track of its growth, remove contaminated vessels, make sure the temperature is right, and move vessels to the next stage when ready.

	SPAWN RUN	PRIMORDIA FORMATION	FRUITBODY DEVELOPMENT
TEMPERATURE	21-27 (30)°C 70-80 (85)°F	(7) 10-21 (32)°C (45) 50-70 (90)°F	(7) 10-24 (>30)°C (45) 50-75 (>85)°F
RELATIVE HUMIDITY	(80) 85-95 (100)%	(80) 95-100%	(60) 80-90 (95)%
LIGHT	n/a but some not detrimental for most strains. (50-100) lux	(4-8) +/-12 hr/day (100-200) 1,000-1,500 lux	+/-12 hr/day (100-200) 1,000-1,500 (2,000) lux
CO_2 PPM, FRESH AIR EXCHANGES (FAE)	5,000-20,000 (50,000) ppm (0) 1 (2) FAE/hr	(400) 500-2,000 (20,000-40,000) ppm (0-2) 4-8 FAE/hr	<1,000-2,000 ppm (2) 4-8 FAE/hr
DURATION	(1) 2-8 (14) weeks	(2) 3-12 (28) days	(3) 5-8 (28) days

▲ Table of environmental conditions. These values are generalized for most commonly cultivated species. Outliers' values are indicated in parentheses. These are ideal values; certain strains may reach beyond them.

FRUITING SPACE: FACTORS TO CONSIDER

Most wild mushrooms fruit in the autumn, when temperatures drop, rains come, and day length is 12 hours or less. In creating a fruiting space, strive to mimic the conditions that elicit fruitings of your particular mushroom(s) in the wild. Whatever the size of your fruiting space, there are four main conditions conducive to primordia formation (pinning) and mushroom development (fruiting) for most mushroom species:

- Reduced temperature
- High *relative humidity* (RH), +/- 85%
- Increased *fresh air exchanges* (FAE), (reduced carbon dioxide/increased oxygen)
- Gentle light (diffuse natural, LED, or fluorescent) for about 8–12 hours/day

Blocks or vessels of substrate need more space in a fruiting room than they do during spawn run, as mushrooms need room to grow into. As with the other fungal living spaces, it is important to keep the fruiting spaces clean, but less meticulously so than the lab and spawn run spaces. Contamination eventually happens during fruiting, and it is up to the grower to decide when potential yields are no longer worth the inevitable spread of contamination that will result from leaving it in the space. When an outbreak of contamination happens, it is important to clean thoroughly to prevent it from spiraling out of control. Increasing FAE in the chamber usually helps decrease contamination rates. The biggest challenge is balancing the needs for high humidity and lots of fresh air.

CONTAMINANTS IN THE HOME

In-home growing can sometimes promote incidental growth of molds and the pollution of the indoor environment. Fortunately, most common contaminant molds are soil fungi that aren't the same as the molds that often grow in houses, but the added humidity from indoor fruiting could encourage black molds and other types in the space. Additionally, spores, especially oyster, can be abundant and harmful to respiratory health. Oyster mushrooms have also been known to take hold in surprising places. These risks can be managed with a good cleaning regimen, good technique throughout, keeping humidity contained, and harvesting mushrooms before they are overgrown and heavily sporulating. The use of a dehumidifier outside the fruiting space can help with this as well.

▲ Oysters fruit from the floor of the fruiting room at Mycality Mushrooms in Arcata, California. WILLOUGHBY AREVALO

▸ Shiitake fruiting in a humidity tent made from snow fencing, a plant saucer, and a bag. JASON LEANE

> **SET AND FORGET TENT**
> Once I had an oyster bag pinning right before I left town for a week. I didn't want to lose the potential crop, so I put it in a humidity tent with thoroughly moistened peat in the bottom. I used a big bag without any ventilation holes to prevent moisture loss, and instead I stuck a living, leafy branch of a shrub into the bag and tightly closed the bag around it. When I came home there was a big flush of perfectly formed mushrooms! The closed system worked because there was ample moisture evaporating from the peat, and the plant and mushrooms exchanged gases there within the bag, preventing mushroom deformation caused by elevated CO_2 levels.

OPTIONS FOR HOME-SCALE FRUITING CHAMBERS

Fruiting in Vessel

Simply fill vessels partway with substrate, leaving closed-off air space at the top. The humidity of the substrate usually provides enough for fruiting. This doesn't work for all species, but reishi, pioppino, lion's mane, enoki, and others with high CO_2 tolerance can fruit like this. Others will either refuse to fruit or will form tough, exaggerated stems.

Humidity Tent

A humidity tent is the easiest and cheapest option, but it has a small capacity and doesn't allow any automation. It is simply a light frame or rack larger than the vessel to be fruited with a plastic bag over it. It can be hung or placed on a surface. It can have a moisture-holding medium in the bottom such as perlite, coir, peat, or a folded towel to evaporate moisture into the chamber. Mist the inside walls of the tent 1–5 times per day. Make sure there are holes or slits at top and bottom to allow airflow and drainage. The heavier CO_2 will fall out the bottom and be passively replaced by fresh air through the top. Having multiple tents can allow blocks of different species and ages to each have their own microclimate, and to segregate contaminated kits that are fruiting anyway.

Shotgun Fruiting Chamber (SGFC)

Invented by Marc R. Keith/RogerRabbit and widely used as a small fruiting chamber, the SGFC uses natural air currents to provide high RH and FAE. Moistened and drained perlite fills the bottom and provides constant evaporative humidity and airflow. As humid air is less dense than dry air, the evaporation causes an updraft and further evaporation as fresh air is pulled in from the holes in the bottom. Supplemental misting is usually necessary, and if mushrooms show signs of CO_2 overload, fanning out stale air with the lid occasionally helps.

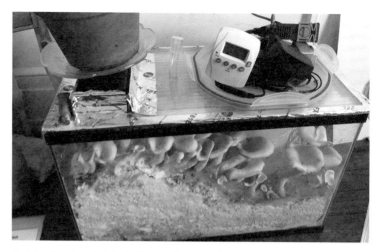

▲ A local oyster strain fruits from coffee grounds in a fully automated, mono-tub-like terrarium. The timer runs a small computer fan intake, and a humidistat activates a pond fogger in-line after the intake. The plant sits atop a vacuum cleaner HEPA filter repurposed as an exhaust filter to prevent contaminating the human living space with spores. CHARLIE ALLER

1. Get a large (+/– 64qt [61 L]), clear-plastic tub with a lid and a bag of perlite (available at garden stores).
2. Drill ¼" holes in a 2" (5 cm) grid on all sides, including lid and bottom.
3. In a large colander, thoroughly rinse and drain perlite in batches, filling tub until 4–5" deep.
4. Situate tub on a table or shelf, not on the floor (to avoid contaminants). Elevate tub at least 2" (5 cm) off its resting surface with blocks and keep sides of tub at least 2" (5 cm) from any walls to allow airflow.
5. Place fruiting blocks or jars within, allowing ample space for mushroom development.
6. Mist 1–3 times per day, or as needed.
7. If mushrooms are too stemmy, increase FAE by fanning or place a fan in the room blowing near but not directly on the tub.

Monotub

Another contribution from the online mushroom community, this system was designed for compost-loving mushrooms, but it can also work for reishi, king oyster, and others that are willing to top-fruit. The optional lining is to prevent primordia from forming on the sides. Pasteurized substrate and spawn are mixed, incubated, and fruited in the tub. Maintenance is minimal if set up properly. Many sizes and designs exist; here is how to make one time-tested model:

1. Get a large (+/- 64qt), clear-plastic tub with a lid, some 2" tape (non-porous), a trash bag or plastic sheet (optional), and some polyfill.
2. With a 1½" (38 mm) hole saw, drill two evenly spaced holes 5" (13 cm) up from the bottom of the tub on each of the long sides. Drill two more 1½" (38 mm) holes just under each of the handles on the short sides. When drilling, use light pressure so you don't crack the tub.
3. Invert tub and wrap the bottom like a gift, taping folds into place. If you wish, you could slip off the wrapping and put it inside tub as a liner, taping the top edge of the plastic to the tub.
4. Fill with hydrated and pasteurized substrate and spawn (ratio depends on mushroom species and substrate) to 1" (2.5 cm) below the lower holes, or less if you plan on casing (casing is explained in Chapter 7). Mix thoroughly and level, but do not pack.
5. Using scissors, cut away excess plastic liner (if using) at substrate level, or above if casing later.
6. Put the lid on the tub and cover the six holes with tape. The lid isn't airtight and allows for enough gas exchange. Incubate.
7. If applying a casing, do so as soon as substrate is fully myceliated, and incubate again.
8. When fully myceliated, initiate fruiting by removing the tape and replacing with a wad of polyfill in each hole. Stuff top holes loosely and lower holes tightly. Adjust polyfill density to dial in RH:FAE balance. Mist and or fan with lid only if necessary. Monotubs can be fruited indoors, or outdoors if temperatures are fairly consistent and agreeable.
9. Substrate block can be taken out of tub to facilitate harvest. Replace substrate block, briefly soak, and drain for subsequent flushes.

▲ Pink oysters cozy up beside bath toys at our house.
WILLOUGHBY AREVALO

The Shower

The bathroom is usually the most humid room in the house, though kitchens often get quite steamy. Hang bags in the shower but outside of the direct spray and shampoo splash zone (mushrooms don't need shampoo, possibly with the exception of lion's mane). A few showers per day may provide sufficient humidity to support fruitings of oysters and others that don't need consistently super-high humidity. And then your mushrooms get to see you naked, which may cause them to grow extra big. If showering alone doesn't provide enough humidity, you can always mist by hand too. A spare shower stall is easily converted to a dedicated fruiting chamber with the addition of racks and a humidifier.

Mini-Greenhouse/Martha

Basically, this setup is a big humidity tent over a rack, automated with humidifiers, lights, and fans. It can be situated indoors or in a shady outdoor location, temperatures permitting. In the online forums, this unit is called a *Martha*, because the first ones were made from Kmart Martha Stewart storage closets retrofitted with racks.

Prefab units designed for starting seeds indoors or growing microgreens work great, but it is also easy to tent your own rack using clear poly sheeting and sheathing tape. A gabled roof helps to prevent pockets of stale air. You will need a way to open the front, so use Velcro, magnets, or clips to make a door. Create a false floor with a big plastic tray or other washable material. Cut a drain hole to release excess water into a pan below. Fill the floor tray with perlite, which will wick up and evaporate condensation that drips down into it. Empty shelf space can also hold trays of moist perlite to raise RH and allow for more FAE.

Place a cool mist humidifier (impeller type) on the false floor and point it to the side so it doesn't blast the bottom shelf. Use a multi-function timer to control its output. You will have to experiment to find what works best for your setup, but start with running a

▲ A mini-greenhouse set up with two cool mist humidifiers, outdoors at Wildwood Ecology Labs. THEO ROSENFELD

cycle of 2–5 minutes on and 6–10 minutes off. Some growers also use an ultrasonic humidifier in addition to the cool mist, but this may be unnecessary unless the ambient humidity is extremely low. Some growers choose to house humidifiers externally and pipe in the mist, but this can reduce the life of the humidifier. Humidifiers can be modified by adding large auxiliary water reservoirs.

For light, situate your mini-greenhouse in a place with diffuse natural light and/or supplement with (compact) fluorescent light mounted outside the tent. Cut slits in the plastic at various levels as needed to increase FAE. A 4" computer fan can blow air out if more exchange is needed. For heat, an oil-filled radiator can be placed near the unit.

▶ A closet fruiting space. Light is provided by T5 fluorescents mounted on the door. Door is opened regularly to give FAE. MAX KIRCHGASSER

Fruiting Room

Many spaces can be retrofitted to be a fruiting room, including closets, pantries, spare rooms, basements, bathrooms, porches, sheds, trailers, greenhouses, shipping containers, barns, and garages. Great care should be taken to not allow water to damage the building itself. Floors, walls, and ceilings should be waterproof and can be painted with marine enamel or epoxy-plastic-based paint. Smooth cement floors with drains are ideal. Heavy-duty sheeting should be used for lining wood floors, and 2–4 mil plastic sheeting meticulously joined with sheathing tape can be used to line the walls and ceiling, as leaks will allow mold growth between the plastic and the walls. Exterior walls must be insulated to prevent the formation of condensation between the wall and the vapor barrier. Alternately, a greenhouse framed with PVC can be built within a room, leaving free space all around it.

Racks made of metal or plastic are preferable to wood, as they will not be readily inhabited by contaminant fungi and are more easily cleaned. Avoid solid shelves because they can hold standing water. Get or build racks that match the type(s) of fruiting vessels you use. If you are hanging oyster columns (strawsages), metal hooks are sufficient. Lidded buckets or bins can be stacked and may not need any racks. Sawdust blocks grown in bags can be supported by racks made with two pipes of metal conduit spaced 4" (10 cm) apart, which is ideal for all-over fruiters like shiitake.

Greater environmental control is needed in larger spaces, so ventilation, heating/cooling, lighting, and humidifiers are usually necessary. It is likely that there will be different microclimates within a fruiting room, so use this to your advantage and place species accordingly. Growing seasonally appropriate strains and species allows year-round production and greatly reduces energy inputs from climate control. Keep in mind that the metabolism of the mycelium contributes heat to the space as well.

▲ Charliceps drops his spore mask to gawk at a huge king oyster in the PVC-framed home basement fruiting room at Mush Luv in Charlottesville, Virginia. NINA O'MALLEY

ENVIRONMENTAL CONTROL IN FRUITING SPACES

Temperature

One of the most important environmental controls is maintaining consistent temperature; this is much easier in a well-insulated space. Heat can be added with an electric baseboard heater or an oil-filled radiator (not too close to any blocks). Heaters are inherently drying and increase the need for humidification. Warm mist humidifiers raise temperature and contribute humidity. Thermostats designed for greenhouses hold up to humid conditions. For cooling, fresh air is usually sufficient, except in hot conditions. In this case, put intake fans on at night and early morning to get the coolest fresh air possible. Air conditioning can also be used, but is financially and environmentally expensive.

Humidification Option	Applications	Pros	Cons	Remarks
Spray Bottle	Humidity tent, monotub, SGFC	Intimate, focused, controllable	Low output, labor intensive, coarse mist	Zep brand recommended for its fine mist and higher output
Pump Sprayer	Humidity tent, monotub, SGFC, Marthas, outdoor beds, logs	Better output, higher-end units make fine mist	Labor intensive, cheap ones make coarse mist	Don't use if it previously contained herbicide, fungicide, or pesticide
Passive Evaporation	Humidity tent, monotub, SGFC, Marthas, greenhouse	Easy and cheap or free	Provides limited humidity	Ideal for off-grid spaces
Hose with atomizer nozzle	Fruiting rooms, greenhouses, outdoor beds, logs	Simple, inexpensive, high output, no refilling necessary	Cheaper atomizers make coarse mist	Can be automated with irrigation timer
Cool Mist (Impeller Type) Humidifier	Marthas, small fruiting rooms & greenhouses	Abundant fine mist	Must be cleaned regularly, vaporizes microbes in reservoir	Ideal for small spaces. Can be automated with multi-function timer and/or auxiliary reservoir
Warm Mist Humidifier/ Vaporizer	Marthas, small fruiting rooms & greenhouses	Adds heat when needed, doesn't vaporize molds or bacteria from reservoir	Must be cleaned regularly	Suitable as a non-drying heat source, but not as sole humidifier
Ultrasonic Humidifier (Pond Fogger)	Marthas, fruiting rooms & greenhouses	Abundant, superfine mist	Must be cleaned regularly, may be pricy	Multi-headed pond foggers can be put in large reservoirs for larger spaces. Duct in fan-forced air to move mist around space
Swamp Cooler (Evaporative Cooler)	Fruiting rooms, greenhouses	Abundant mist, cools space, easy to build	Must be cleaned regularly, may cause excessive water buildup	An internet search for "DIY swamp cooler" yields numerous designs

Humidification

Maintaining high humidity is of utmost importance for fruiting. Humidifier reservoirs must be cleaned regularly (weekly). Hydrogen peroxide can be added to the water in a humidifier to reduce microbial buildup. Purified water is recommended; some cultivators boil and cool tap water before use. Mineral buildup is also an issue in humidifiers, especially with hard water. Deposits can be cleaned with vinegar, soda pop containing phosphoric acid, or a product called CLR. Most humidifiers have filters that must be changed every so often.

Ventilation

Fruiting rooms can be set up with an intake fan for positive pressure or an outtake fan for negative pressure. Each has its pros and cons. Inside a home, where preventing water damage is important, negative pressure is preferred—moist air is vented to the exterior, and fresh air is drawn in from the rest of the house. Positive pressure would push humid air out any gaps or leaks in the room and into the house. A gentle, oscillating fan within the space helps to circulate air; just be careful to position it so it doesn't blast and dry out your blocks.

Light

Fruiting rooms need a moderate amount of light in the warmth range of natural sunlight—basically, enough to read by. Diffuse sunlight, CFLs, fluorescent tubes, and LEDs are all good options. Whatever artificial light you use, put them on a 12hr-on/12hr-off cycle on a dual-function timer. It need not be in sync with day and night if the room isn't getting natural light; it's better to have the lights in sync with when you will tend them. Unlike flowering plants grown indoors on a strict 12/12 cycle, light during their dark cycle won't cause significant problems.

Initiating and Fruiting in a Controlled Environment

Initiating and fruiting can be two of the most challenging parts of the process. Setting up your fruiting space well is key. To coax mushrooms from mycelium, the cultivator must learn to recognize

▲ A DIY pond fogger—a 10-head ultrasonic hydroponic fogger on a float in a trash can. Fan-forced air ducted in the top pushes mist out a hole in the side. DALLAS LAWLOR

▲ These oysters are lavishing in the pink glow of LED grow lights at Northside Fungi. WILLOUGHBY AREVALO

▲ The outtake fan at in the basement fruiting room at Northside Fungi in Enderby, BC. Note the spore buildup. WILLOUGHBY AREVALO

MONITORING AND AUTOMATING ENVIRONMENTAL CONTROL

Hygrometers measure the relative humidity (RH) of the air. Placing one in your fruiting chamber can be crucial for dialing in automation of humidifiers and FAE, especially if it is a larger space. Reasonably priced digital hygrometers lose accuracy at around 80% RH, so an analog unit, available at cigar shops, is far better for fruiting rooms. To calibrate, wrap in a wet towel for an hour, then use the screw on the back to calibrate to 99%. Calibrate weekly or as needed.

Thermometers are equally important in both incubation and fruiting spaces. The most useful are digital models that record min/max temperatures.

Lights, fans, humidifiers, and heaters can all be run through dual-function or multi-function timers; these allow for manual adjustment of automatic functioning. Many models exist for automated environmental controllers with thermostats, humidistats, and even CO_2 detectors, but they are not cheap. For the tech-savvy, DIY controllers can be made using a Raspberry Pi or Arduino.

▲ A DIY environmental sensor for remote monitoring in the fruiting shed at All the Mushrooms in Powell River, BC. JASON LEANE

Work Spaces, Tools, and Equipment • 49

▲ These nameko self-initiated inside the jar. The pins are only 5 mm tall.
WILLOUGHBY AREVALO

when the fruiting conditions need adjusting. Humidity must be very high for pinning, and should remain quite high through the fruiting period. Appropriate levels of fresh air, light, and warmth are also necessary. The following table will help guide you to fine-tuning your environment. Species profiles in Appendix 1 give more info for various mushrooms.

Troubleshooting Fruiting

PROBLEM(S)	POTENTIAL CAUSES	SOLUTIONS
Pins drying out/ caps cracked	RH too low, wind too harsh, light too intense	Increase RH, shelter from wind, decrease light intensity, remove aborted pins
Primordia never form	RH too low, CO_2 too high, temp too high or low, substrate unable to support fruitings, improper fruiting surface	Adjust conditions to suit your species/strain. Cold shock by refrigeration or cold water dunk. Try again with reliable substrate
Elongated stalks	CO_2 too high, inadequate light	Increase FAE, increase light
Poorly formed caps	CO_2 too high, genetic mutation, nutrient deficiency, viral disease	Increase FAE, try different culture source, try different strain
Waterlogged mushrooms	RH too high	Reduce misting, increase FAE, allow harvested mushrooms to dry partially in fridge
Bacterial blotch on caps	Bacterial contamination in fruiting space, too much water on caps	Clean fruiting space, adjust/reduce misting
Pallid mushrooms/ spiraling stalks	Inadequate light	Increase light
Heavy spore load in fruiting space	Overmature mushrooms	Harvest earlier
Heavy pin set but many pins abort	Too much fruiting surface, RH dropped too low	Remove aborted pins, reduce fruiting surface, keep humidity high

OTHER SPACES

Garden and Greenhouse

In most climates, there is at least part of the year when mushrooms can be fruited outdoors with little to no environmental control, especially when sited in shady and humid microclimates. Fruiting vessels can be nestled under hedges, shrubs, and garden veggies, or even partially buried in the ground. Humidity tents can be hung in low branches or placed on balconies or shady sides of buildings. Shady areas of greenhouses often have ample humidity, oxygen, and light, especially when plants are also being grown. Ensure that temperatures are compatible with the mushroom strains you put in there.

Fungus gnats abound in greenhouse soils, so put sticky traps near your mushrooms. For growing mushrooms directly in the garden, see Chapter 8.

Substrate Storage and Processing

Most fruiting substrates are cheaper and easier to obtain in larger quantities, so if you have room to store them, that is ideal. Straw should be kept dry, but sawdust can get rained on and aged for better water retention and therefore better yields. Keep it out of direct soil contact to keep it from harboring tons of contaminants, but it will be pasteurized or sterilized before you use it, anyway. Grains are best stored inside to prevent rodents from feasting. Straw shredding, sawdust hydrating, log drilling, and substrate mixing can be messy, so you will need a place where you can easily sweep up or where mess doesn't matter.

Compost

When possible, composting spent substrates at home is a rewarding practice—the resulting compost is an awesome soil amendment for garden plants. Compost piles are also breeding grounds for *Trichoderma* (which is great for plants) and other contaminants, not to mention slugs and gnats, so if possible, put some space between compost and the rest of your mushroom enterprise.

Overall Layout and Workflow

While most home growers are limited by the constraints of their spaces, it is best to respect proper flow of materials and caretakers through the grow spaces whenever possible (see "Vectors of Contamination: The Cultivator," in Chapter 4). The spatial organization of the various spaces can make a big difference in reducing contamination. Materials should never have to travel backward through the different spaces. When equipment, such as reusable cropping containers cycle back through, they should first be cleaned and sanitized.

▲ Shady areas in greenhouses used for growing plants make good fruiting spaces if not too hot. WILLOUGHBY AREVALO

MY CURRENT SETUP

I live with my wife (who has high standards for cleanliness and organization) and our toddler (who is good at getting into everything within reach) in a 700-square-foot, one-bedroom basement suite with a partially fenced and mostly shady backyard that we share with the neighbors/owners. I do my sterilization, grain prep, and liquid culture lab work in the kitchen. In the rare case I need a still air box, I borrow one from a friend. I incubate/do spawn runs in the water heater/furnace closet, store LCs on top of the fridge, and do some bulk inoculations at the dining table. I store tools, supplies, and spawn run overflow in a cabinet and on a shelf in the laundry room. I use the bathroom or laundry room for fruiting when outdoor temperatures are not conducive.

Outdoors, I store my giant pots and burners, my steam pasteurizer, sawdust, and straw. I shred and ferment straw; mix, moisten, pasteurize, and inoculate fruiting substrates; load vessels; drill and put silicone on jar lids; inoculate, incubate and fruit logs; incubate naturalized woodchip spawn in bins or burlap sacks, and compost and vermicompost spent and contaminated substrates along with kitchen and garden scraps. I do most of my fruiting outside, in humidity tents/mini-greenhouses nestled into shrubby and shady zones where plant transpiration provides a decent baseline of RH and O_2. When weather is conducive, I remove the plastic coverings and let the clouds and rain do the work. Because I live in the rainy PNW, and because I keep lots of decomposing organic matter around and mulch heavily, I have a high population of slugs. So I hang several humidity tents from shrubs to prevent the slugs from decimating my yields. I mist my fruiting chambers by hand, with the mist nozzle on a hose.

▲ Humidity tents hang from a shrub in my garden in May. WILLOUGHBY AREVALO

SANITATION AND TECHNIQUES TO AVOID CONTAMINATION

▶ Lion's mane fruits from grain despite horrid mold contamination. DANIELLE STEVENSON

TO GROW MYCELIUM in culture, one must take many steps to ensure that it is always given the advantage over competing organisms. We use heat, chemicals, or other treatments to sterilize or pasteurize substrates and to clean tools and workspaces (see Chapter 7 for details on substrate treatments). We filter and limit airflow; work with quick, precise movements; and we banish infected mycelium and substrates to the compost pile.

It seems paradoxical. Mushrooms grow in the dirt, right? So why do we need to be so careful to prevent contamination? In natural systems, many fungi will start to grow, but few survive long enough to fruit. Most are overtaken by parasitic fungi, eaten by bugs, infected with viruses or bacteria, and/or suppressed by competitors vying for the same food source. Much of their energy is put toward self-defense. In culture, the mycelium is isolated from other organisms. With an abundance of nutritious food, its focus shifts from its immune system to its digestive system. The mycelium produces only the chemicals it needs, so it doesn't bother producing defensive chemicals; so, when an antagonistic organism finds its way into a pure culture, the mycelium's defenses are down, and it more easily succumbs to attack. Eventually, its immune system responds and may overtake the invaders, but sometimes it is too late.

> Air is one of the most important and challenging vectors to manage. *Airport lids* are jar lids modified for liquid culture use, fitted with an air filter and a *self-healing injection port* (SHIP). With an airport lid, one can use a needle and syringe to move liquid culture around from jar to jar in the open air with only minimal risk of contamination by airborne microbes. This online-forum-born technology has revolutionized home mushroom growing by eliminating the need for a clean air space (glove box or flow hood) for culture work and the creation of grain spawn. I do most of my lab work on my kitchen counter—even though kitchens tend to be full of microbes. The airport lids let me get away with this. Because they allow containers to stay securely closed, they also help to reduce contamination by the cultivator, environment, and pest vectors.

Remember that those tasty, nutritious substrates we prepare for our fungal companions are very appealing to many other types of organisms. For example, bacteria can reproduce through a substrate more quickly than our mycelium can, creating inhospitable conditions for our fungal friends. It's like when we prepare soil for planting—waves of weed seeds germinate and must be weeded out if we want our plants to thrive.

All of this said, fungi are resilient and able to tolerate some level of stress. In fact, a moderate amount of stress often encourages vigor for a complacent mushroom culture. Some cultivators have limitations that do not allow them to do everything possible to nurture their mycelium, while some cut corners out of inexperience or laziness and still have success. Every mushroom cultivation setup I've ever seen, from tiny home grow to big farm, has some amount of contamination present, especially in the fruiting spaces. Sometimes the contamination is minor, and the mushrooms overcome.

VECTORS OF CONTAMINATION AND MANAGEMENT STRATEGIES

There are seven distinct vectors for contamination: air, substrate, tools, inoculum, cultivator, environment, and pests. The most vulnerable moments in the cultivation process are during inoculations—any time we transfer mycelium to a new substrate—so this is when the greatest level of precaution is demanded. When you encounter contamination, trace it to its source. This can be determined by observing where and how the contamination first presents itself.

Air

The air is full of fungal spores, bacteria, viruses, and yeasts. These microbes travel on air currents and fall downward in relatively still air. Anytime we expose a substrate to unfiltered air, we welcome them in to share in our mycelium's meal. Likewise, we must assume that any tools, hands, work surfaces, or containers that are exposed to the open air are covered in microbes. As our mycelium needs oxygen to respire and survive, we must allow some air to pass in and out of our containers, but we must filter this air so that microbes may not enter. Traditional mushroom lab techniques must be done in a still or clean air space, as explained in Chapter 3. If the air vector is the culprit, contamination will typically show up on the surface of the substrate that was exposed to air.

Substrate

The substrates and media that we feed to our mycelium harbors microbes, so we usually have to kill the microbes before we inoculate. Substrate treatments are covered in detail in Chapter 7. When a substrate is not sufficiently treated, contamination tends to appear throughout the substrate or in patches. A way to ensure that your sterilization was effective is to keep a "blank"—one vessel of substrate that is not inoculated but is dated, labeled as blank, and incubated along with the rest of the batch. If your blank gets contaminated, you know that your sterilization was ineffective. Once the rest of the batch grows out okay, the blank can be inoculated as to not waste the substrate and container.

Tools

Any tool that touches the culture, spawn, or treated substrate must also be treated to prevent it from introducing contaminants. Lab tools such as needles and syringes, scalpels, tweezers, and inoculation loops can be wrapped tightly in aluminum foil and sterilized in a pressure canner (PC) for 15 minutes at 15 psi between work sessions. Metal tools can be cooked for a longer time, but polypropylene syringes will melt when sterilized over 30 minutes. Plastic scalpel handles and blade covers are not designed to be re-sterilized and may partially melt in a PC. The working ends of sterile tools should not be allowed to touch any surface that has not also been sterilized or at least sanitized, or be left exposed to the open air.

Since tools are often used repeatedly in a single work session, one will also need to sterilize the working end of a tool between

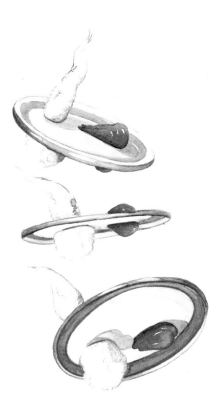

▲ Homemade airport lids featuring a self-healing injection port (SHIP) made with hi-temp RTV silicone and an air filter made of polyfill. CARMEN ELISABETH

▲ Flame sterilizing needle
WILLOUGHBY AREVALO

uses. This is done by heating until red-hot, often in the flame of a clean-burning alcohol lamp.

These are inexpensive to buy or easily made with a small jar and a cotton wick. Propane torches are also commonly used. You can use a candle or lighter to the same effect, though this gets soot on the tools, which is undesirable. Working with open flames near flammable liquids carries a fire risk, so exercise caution. A safe, elegant, and effective (though expensive) option is the Bacti-Cinerator, which is a super-hot electric heater inside a hollow cylinder. You stick the working end of your tool in for just a few seconds and anything alive on it will be sent to the spirit world. Allow hot tools to cool before touching live mycelium with them; this only takes a needle or scalpel blade a few seconds.

If a tool carries in contamination, it will show up first at the spot where the tool touched the substrate.

Inoculum

If the culture or spawn is not pure, then the contamination will present itself in the freshly inoculated substrate. Sometimes culture or spawn doesn't look contaminated even though it is. This is especially true when molds are present, because their mycelium is also white, and especially so in liquid cultures. However, molds turn colors when they produce the spores that enable them to reproduce rapidly and take over. If you suspect that your culture or spawn are contaminated but you aren't sure, either wait and watch how it develops, or run it to a small batch of LC, agar, or grain to see what happens. Using contaminated inoculum typically results in a dramatic and spectacular explosion of contamination in the newly spawned substrate with no hope for recovery. You will see it growing out from points of inoculation all through the substrate. In this case, you will have to work backward to test the purity of the lineage of culture from which it came.

Cultivator

The human body is a locus for diverse and abundant microbial life. Our hands, mouths, noses, breath, hair, and clothes are considerable sources of potential contamination. Our hands are our main tools, yet we cannot ever fully sterilize them, at least not without getting third-degree burns. Therefore, our technique must be deliberate, precise, efficient, and quick. Many cultivators choose to wear nitrile gloves while doing inoculations because they can be sanitized with alcohol more thoroughly than hands, but this is no substitute for good technique. I typically do not wear gloves unless I will be directly touching the spawn and/or substrate. During inoculation, we must avoid placing our hands directly over substrates, touching the working end of tools, the inside of containers, or any surface that will be exposed to the substrate. We must avoid breathing directly over our work and avoid speaking, coughing, or sneezing while doing transfers.

The workflow is of utmost importance. In a single work session, a cultivator should work from cleanest to dirtiest. Begin with any and all lab work, then move things into and manage your incubation space, then do whatever harvesting or fruiting space work you need to do, then do substrate prep, then finish the session by disposing of spent substrate and contaminated vessels. This way, you will not carry microbes backward through the cycle. Cultivator-borne contamination can present in a variety of ways, but it is often inconsistent through a batch and tends to show up at the place where a cultivator touched, breathed, etc.

▲ A polyculture of molds develops on grains that were inoculated with contaminated liquid culture.
JASON LEANE

Environment

The spaces where we work and mycelium and mushrooms grow harbor microbes that may find their way into our work. We need the greatest level of cleanliness in a lab or transfer space, where prepared substrates are inoculated with culture or spawn. Even closed containers can be sometimes breached and become susceptible to contamination, so the incubation space should also be relatively

> When I worked in the lab at Mycality, I always showered and put on clean clothes just before driving to work (in my dirty-ass truck). I had a designated pair of lab-only shoes. I wore gloves for the first months I worked there, but stopped doing so once I developed solid technique, and I had low contamination rates. In some farms, lab workers are required to wear Tyvek suits, gloves, booties, and surgical masks. In the context of home-scale growing, I regard these extreme and arguably unnecessary measures as sterility theatre.

clean. Fruiting spaces, with their high humidity, airflow, exposed mycelium, and sporulating mushrooms, tend to get dirty quickly and should be cleaned regularly. Contaminated mycelium should be removed or quarantined to reduce the spread of contamination. In an ideal situation, surfaces such as worktables, walls, and racks should be made from non-porous materials for easy cleaning and to provide minimal sanctuary for microbes. In reality, people growing mushrooms in their homes have to work with whatever spaces they have and will often need to bend these rules and hope for the best. I have seen (and had) successful mushroom-growing operations in less-than-clean spaces. Environment-borne contamination can be hard to trace, but if you are getting regular contamination coming in unexpectedly, your space may be the culprit.

Pests

Animals that want a bite of our mycelium or substrates can carry in contamination. Fungus gnats are a significant concern in incubation and fruiting spaces, as they both eat our mycelium and disperse spores of many competitor fungi, including pathogenic fungi that may attack our mycelium. Effective sticky traps can be made by coating a yellow surface with Tanglefoot glue; and drowning traps can be made with a bowl of soapy juice or a vinegar solution. *Bacillus thuringiensis* bacteria, often referred to as BT (i.e. Gnatrol), kill fungus gnats and other pests and are non-competitive with mycelium. But the removal or quarantine of gnat-infested mycelium is crucial.

Carnivorous plants are sometimes proposed as controls. While they are beautiful and effective fly controllers, and most thrive in humid environments, unfortunately, they also attract their prey, which may be counterproductive. If feasible, the introduction of green tree frogs into the growing space is a great way to deplete an insect population.

Pleasing fungus beetles are significant pests in some regions, munching spores and laying eggs in mushrooms, riddling them with larvae. These red and black beetles are difficult to get rid of, but

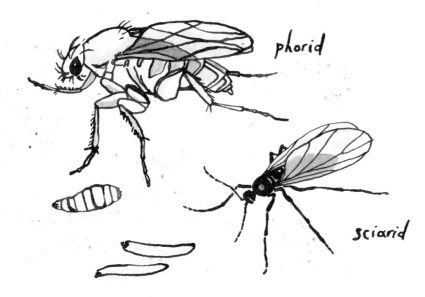

▲ Phorid and sciarid fungus gnats are common in mushroom grow spaces and in soil. CARMEN ELISABETH

infested mushrooms can be tapped over a bucket of bleach solution to kill the adult beetles. Many insect pests are particularly attracted to spores, so harvesting mushrooms before they are fully mature can reduce infestations. Flying insects can be sucked out of the air with a vacuum.

Mycophagous (fungus-eating) mites can be very problematic. A variety of species of tiny mites can invade mushroom grows, munching on and reproducing through cultures, spawn, fruiting blocks, and dried mushrooms. Beware of leaving old dried mushrooms lying around on mantles, dashboards, altars, and shelves. Both wild and cultivated specimens can develop serious mite infestations that can wreak havoc on a cultivation space. Management is difficult and involves thorough and repeated cleaning of the space, sticky mat barriers for storage of cultures, disposal of infested materials, and the use of heat to kill mites in tools and equipment.

Rodents are attracted to grain spawn and can smell it through plastic. Rodents are also attracted to stored substrate materials

▲ Grain contaminated by wet-spot bacteria. WILLOUGHBY AREVALO

such as straw, grain, and bran, so keep these in sealed containers. Prevention by closing access holes is advised, and snap traps are a good remedy. Contamination by pests is usually evidenced by chew marks, feces, or larvae.

COMMON CONTAMINANTS: RECOGNITION AND MANAGEMENT

Different microorganisms can show up in our work at various parts of the process. To manage contamination, you need to be able to recognize it. The more simple sugars and nitrogen available, the more susceptible the substrate is, making liquid or agar culture media the most vulnerable, followed by grain, nutrified sawdust, and bulk substrates, respectively.

Bacteria

Bacterial contamination can occur in culture media, grain, or substrates, especially when overhydrated. To prevent problems with bacteria, avoid having any standing water at the bottom of a vessel or oversupplementing your solid substrates. Oversupplementation can cause high temperatures to result from the metabolism of the mycelium, favoring thermophilic (heat-loving) bacteria. Whenever oxygen cannot reach the center of the mass of substrate, conditions can become anaerobic, suffocating mycelium and promoting the growth of bacteria. Bacterial growth often produces chemicals in the substrate that are toxic to mycelium, further inhibiting its growth. Masses of substrate that are too thick, too moist, or too tightly packed are prone to anaerobic conditions. Bacterial contamination in solid substrates tends to look wet or greasy, and is often in blotchy or patchy patterns. This is especially evident in grain, where it is called *wet-spot*. *Bacillus subtilis* is often the culprit, as it easily forms endospores to withstand extreme conditions (i.e., heat). It is ubiquitous in soil and our guts, and is not dangerous to us.

Molds

Molds are typically the most common and problematic contaminants in mushroom growing, and they can appear in any stage of the process. Most have white mycelium that can be mistaken for mushroom mycelium until they sporulate and turn colors. These spores cling to surfaces, hands, and pests, from which they spread easily. By this time, they pose a significant risk of multiplying exponentially, so moldy substrates should usually be disposed of as soon as they are detected. In minor cases, molds can be suppressed with a spray or dab of 3% hydrogen peroxide, which will kill their spores but not their mycelium.

Many different molds like to crash a mycelium party, especially when FAE is too low. They can sometimes be recognized to genus by the color of their spores, though certain identification usually requires a microscope. *Penicillium* spp. are blue, grey-blue, or blue-green. *Mucor* spp. are referred to as black pin mold because of their appearance. *Aspergillus* molds can be black, yellow, green, or even pinkish depending on the species. Some, such as the yellow-green *A. flavus*, produce strongly carcinogenic aflatoxins and should be disposed of with care. *Penicillium*, *Mucor*, and *Aspergillus* species show up most often during the culture or grain spawn stages but may also be found on fruiting substrates.

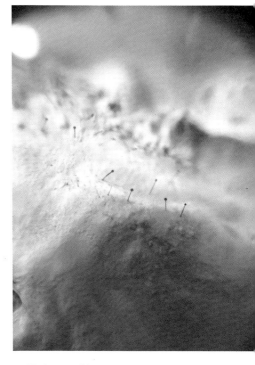

▲ Black pin mold appearing on the surface of old, overgrown grain spawn. WILLOUGHBY AREVALO

Most molds are simply competing for food, while others are pathogens of the mushroom mycelium. Either can significantly reduce yields, but the pathogens are particularly impactful. The most notorious of the pathogenic molds are members of the genus *Trichoderma*. These so-called *forest green molds* are ubiquitous in soil, supporting plant health by attacking fungi that are plant pathogens and exuding powerful enzymes that speed the decomposition of organic matter and formation of humus. They can attack mushroom mycelium in pretty much any stage of cultivation but are especially prevalent in fruiting substrates, causing them to become soft. Their mycelium is white before it produces green, blue-green,

or grey-green conidiospores. If small patches of *Trichoderma* erupt while mushrooms are in the fruiting process, mushrooms may still be able to develop normally and be safe to eat, though yield may be reduced. But don't eat moldy mushrooms. Substrates heavily contaminated with *Trichoderma* should be composted far away from the growing areas, or disposed of in municipal compost bins. Sanitize exposed surfaces after an outbreak.

Slime molds show up sometimes in outdoor fruiting spaces, beds, and logs, particularly when large numbers of bacteria are present. These highly intelligent organisms are often really cool-looking and are usually non-competitive with mushroom mycelium.

Yeasts

Though not often discussed in mushroom cultivation literature, yeasts are sometimes present in culture media and substrates, consuming sugars and oxygen, and causing the production of alcohol and carbon dioxide. Without microscopic analysis, they are hard to differentiate from bacteria, though they can sometimes be noticed by their characteristic smell. I have most commonly encountered them in cloudy, contaminated liquid media into which wild mushroom tissue was cultured.

▲ Green mold, probably *Trichoderma* spp., growing on sterilized, nutrified sawdust. It was introduced by impure liquid culture. WILLOUGHBY AREVALO

▲ *Trichoderma* sporulating from the end of a shiitake log just 6 months after inoculation. WILLOUGHBY AREVALO

Mushrooms

Other species of mushrooms can compete for substrates and may sometimes appear unexpectedly from straw, woodchips, and logs. These can reduce or eliminate the yields of your intended mushrooms. Beautiful little inky caps as well as grotesque cup fungi are common inhabitants of straw. If they show up, it probably indicates that your straw was insufficiently treated prior to inoculation. Outdoor woodchip beds may harbor significant populations of weed fungi such as *Stropharia ambigua*, *Agrocybe praecox*, *Tubaria* spp., *Psilocybe cyanescens*, *Leratiomyces ceres*, *Hypholoma* spp., various bird's nest fungi, and others.

Logs will often support the growth of mushrooms other than those we introduced; common competitors on logs include *Stereum* spp., *Trametes versicolor, Hypoxylon* spp., *Daldinia concentrica, Xylaria hypoxylon, Schizophyllum commune,* various crust fungi, and jelly fungi.

These fungi may have already been present when the wood was harvested. We could consider them uninvited guests, but they might consider the mycelium we inoculate with to be the uninvited guest. Some of these competitor mushrooms are medicinal and make a decent consolation prize. Unless the wood is fully covered in these other mushrooms, there is still a chance that our mycelium will fruit, albeit less than it would without competitors.

Viruses

Being much smaller than bacteria and fungal spores, viral bodies may be able to breach our air filters and infect our mycelium. They can be the cause of slow or poor mycelial growth and the decline of older mushroom cultures. They are hard to detect, but they may cause deformities in fruiting bodies. If a viral infection is suspected, I recommend going back to an earlier generation of the culture if possible, growing it out alongside the culture showing symptoms of viral depression, and comparing their growth rates and fruiting body formation.

STARTING AND MAINTAINING CULTURES

GET CULTURED

Cultures are starter mycelium grown on a nutrified medium, either liquid or agar based. Besides making your own (see instructions below), cultures are available from spawn suppliers, culture collections, online vendors, and other cultivators (see Appendix 2). Most mushroom growers love to sell, trade, or share cultures, so make some friends and get cultured!

Spore Prints

New cultures can also be started from spores, so below are instructions for starting liquid and agar cultures from spores. This is just one of many ways to make a spore print; this one is optimized for LC users without an aseptic transfer space. Prints can also be made on aluminum foil, glass, or clean printer paper. Spore prints are often contaminated with bacteria, so expect some margin of error and try several times if needed. You'll need: a fresh mature mushroom (wild or cultivated, and as clean as you can keep it), hydrogen peroxide, cotton balls, alcohol spray, clean scissors, and a new zip bag.

1. Sanitize hands or gloves with alcohol.
2. Clean top of cap carefully with a cotton ball dipped in hydrogen peroxide, being careful not to touch the gills.

◁ Sanitizing cap surface. ISABELLE KIROUAC

◁ Placing cap in bag. ISABELLE KIROUAC

3. Sanitize scissors with alcohol and cut stalk off mushroom.
4. Sanitize hands or gloves again.
5. Holding sideways, open zip bag. Place cap gills-down inside and seal.
6. Let rest approximately 12 hours or until a heavy deposit of spores is present.

7. Sanitize hands/gloves and hold bag zip side down, holding cap in place. Open bag, letting cap fall out. If the bag is wet with condensation, hang it upside-down with a clothespin until dry. Reseal and label with species, strain/location of harvest and date.
8. Store in fridge, or cool dry place. Spore prints can remain viable for years.

This productive reishi strain was cloned by Olga Tzogas from a mushroom she collected in a city park in Rochester, New York. OLGA TZOGAS

THE IMPORTANCE OF GENETICS

Cultivators refer to a unique dikaryotic (mated) mycelium as a *strain*. Most known strains have names and/or numbers, and some are catalogued in large collections such as the American Type Culture Collection (ATCC). Because of the genetic variability of spores, mushrooms can adapt very quickly from generation to generation, and different qualities may evolve within a given species in different habitats or parts of the world. There may be significant differences between one strain and another of the same species, even if they came from spores from the same mushroom, but especially if they came from different continents, habitats, or substrates. You could germinate multiple spores and isolate new strains from the resulting dikaryons, but for most purposes it is easier and more consistently effective to work with cloned mycelium from a known strain. A particular strain may be adapted to a novel substrate, such as shiitake #034 that will fruit from straw. Many strains have been tested for their fruiting temperature ranges; so, for example, one may grow warm-weather strains of shiitake in the summer and cold-weather strains in the winter with minimal or no heating or cooling of the fruiting space.

We can bring wild strains into cultivation by cloning mushrooms that we find growing in their natural habitat. These strains will often be well adapted to the local climate and substrates and allow us to boost the populations of native mushrooms from our bioregions. Likely most mushroom strains in cultivation worldwide originated as clones of wild mushrooms, native to some part of the world. Commercial strains (those used in production farming) have been selected for the quantity and/or quality of fruiting bodies they produce and for other traits. They can easily be obtained by cloning mushrooms from the grocery store. Some wild strains will not be as high yielding as commercial strains, which have been selected by humans rather than by their ecosystem, their fitness being determined by other factors.

▲ Liquid cultures. L to R: wine cap, pioppino, elm oyster.　WILLOUGHBY AREVALO

LIQUID CULTURE

I love liquid culture (LC) because it is beautiful, cheap, easy, fast-growing, vigorous, versatile, and I don't need a clean lab space to make and use it. Liquid culture practices (technically, *submerged fermentation*) have been used since soon after pure culture techniques were developed about a hundred years ago, but their use was only applicable to the home grower since the last couple of decades. Most cultivation books that describe it offer a method that I think of as liquefied culture—pure agar cultures or grain spawn are put in a sterilized blender with sterile water under aseptic conditions, then slurried and incubated before being used to inoculate grain spawn. Our method grows mycelium directly in nutrified water, in jars fitted with self-healing injection ports (SHIPs), allowing the transfer of mycelium from jar to jar with a hypodermic needle and syringe in the open air. This *airport lid* tek is a boon to home growers and perhaps the greatest innovation to ever come out of the online mushroom community.

Starting and Maintaining Cultures • 69

THE ONLINE MUSHROOM COMMUNITY

Since early in the history of the internet, people have been sharing information, insights, and *teks* (techniques) on home-scale mushroom cultivation on online forums such as shroomery.org, mycotopia.net, and mycotek.org. These sites have fostered a major evolution in home-scale cultivation techniques. While the forums and teks were mainly developed by and for magic mushroom growers, much of the information can apply to the cultivation of other mushrooms, and many sites have sections dedicated specifically to gourmet and medicinal mushrooms. Any questions you have about mushroom cultivation has probably already been asked and answered on the forums. The FAQ section is often a good starting place. But, beware: a wide variety of perspectives, approaches, practices, and opinions are represented, not all of which are worth following.

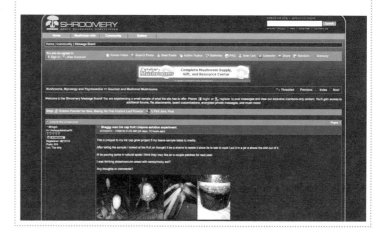

Making Airport Lids

You'll need: a drill with ³⁄₁₆" and ⁵⁄₁₆" bits (or a hammer and nails, in a pinch), a tube of high-temp RTV silicone gasket maker (available at auto parts stores), jar lids (metal or polypropylene), some polyfill for an air filter (for grain or sawdust jars, two layers of micropore tape or a jumbo cotton ball can work, but these are not recommended for LC), and a ventilated place to dry the silicone. I made a drying rack out of small pieces of wood with small gaps between each. I like to wear gloves when applying the silicone.

▲ Making the air filter with polyfill. WILLOUGHBY AREVALO

1. With a board beneath, drill two holes in each lid with a ³⁄₁₆" bit, dividing the lid into thirds. Be careful to not scratch the inside coating with metal shavings, as this leads to rust.
2. Switch to a ⁵⁄₁₆" bit and make one of the two holes in each lid bigger.
3. Use wire cutters to trim any hangnails and gently brush away any shards of metal.
4. To make the SHIP, cut open the silicone tube's nozzle diagonally to give a roughly ³⁄₁₆" hole. On the top of the lid, apply a bead of silicone around the outside of the big hole and spiral inward to fill it, then cleanly smear the tip down on the lid to end the bead.
5. Flip the lid and repeat on the bottom, squeezing a little extra up through the hole to create a nipple over the hole (this helps the needle find the hole), then smear off to end the bead. Avoid making craters or peaks. This takes practice.
6. Position the lids so the silicone isn't touching anything and let dry in a ventilated place for 24 hours or so.
7. To make the air filter, pinch off a little polyfill and twist the end to a point. Thread it up through the small hole from the bottom to the top.
8. Pull through until snug. If it slips through, you need a bigger wad.
9. Gently tease away any excess from the bottom with fingertips. Rub between palms to mat it down a bit. You should be left with a marble-sized wad on the inside of the jar.

Preparation of Liquid Culture Media

Most LC recipes are formulated to have a sugar content of 4% or less by weight. More than this is toxic to mycelium. Other ingredients can be added in small amounts for protein, lipids, starches, minerals, vitamins, and more. Adding a little bit of an infusion or a few particles of the final substrate can speed myceliation during spawn run. Beware that some ingredients cause cloudiness that can disguise contamination. For general purpose, I prefer regular-mouth pint (500 ml) Mason jars. To keep things interesting for your cultures, change up the recipe when expanding them.

You'll need: a gram scale or measuring spoons, measuring cup, jars, airport lids and rings, agitators (marbles, pieces of broken glass, crystals, or magnetic stir bars), big pot, jar funnel, ladle, paper towel or clean cloth, big spoon or whisk, aluminum foil, and a pressure canner (PC) or alternative.

1. Measure out all ingredients. Calculate total volume based on how many jars you need/can fit in your PC. Consider saving room for a plain water jar.
2. Heat water to warm (not boiling) in a pot.
3. Stir in other ingredients until fully dissolved.
4. Using a jar funnel and a measured ladle or pitcher, fill jars not more than halfway.
5. Drop a stir bar or other agitator into each jar.
6. Wipe up any drips from the rim of the jars.
7. Put on airport lids and screw down rings until snug.
8. Cover jar lids with foil to keep filters dry, and load into PC. Include any syringes or other tools in need of sterilization, wrapped in foil.
9. Sterilize for 15–20 minutes at 15 psi (see "Substrate Treatments," in Chapter 7). Longer cooking will caramelize and toxify the sugars and melt the syringes.

SELECT LIQUID MEDIA RECIPES (PER 1000 ML WATER)

Consider these recipes as starting points, and let your creativity run. Avoid sucrose (cane or beet sugar). Complex carbohydrates may lead to more vigorous growth than pure sugars. A dash of gypsum and/or nutritional yeast can be added to any of these recipes for minerals and vitamins, but they will reduce clarity. Measuring by weight is more accurate than by volume. For a link to FastFred's Media Cookbook, see Bibliography.

Malt Extract LC (MELC)
4 Tbsp (40 grams) light malt extract

Malt Extract Dextrose LC (MDLC)
2 Tbsp (20 grams) light malt extract
2 Tbsp (20 grams) dextrose

Honey LC (HLC)
2 Tbsp (40 grams) honey

Sabouraud's Dextrose Broth (SabDex)
4 Tbsp (40 grams) dextrose
1 Tbsp (10 grams) polypeptone or neopeptone
1 L distilled water

Complete LC (CLC)
2 Tbsp (20 grams) light malt extract
2 grams peptone
0.6 grams yeast
10 drops vegetable oil
2 grams ground grain

Grain Cooking Water LC (GCWLC)
Full-strength or diluted water saved from boiling whole grains (i.e., for grain spawn). Use fresh or freeze promptly for later use. Pour through coffee filter to clarify (but some batches will remain cloudy). Should be sterilized along with grain, or for 1–2 hours, as grains harbor many resilient bacterial endospores, and there are few simple sugars to caramelize. I like the quality of growth I get in this medium.

Tissue Culturing to Liquid Culture

This biopsy technique works well with most fleshy mushrooms, though tough or tiny mushrooms can be nearly impossible to catch this way. Waterlogged mushrooms should be allowed to dry out a little in the fridge first. Dry mushrooms can be soaked in 3% hydrogen peroxide to rehydrate. For improved chances, repeat process with multiple jars of media.

You'll need: a mushroom, a sterile syringe fitted with a 16-gauge needle, a sterilized jar of water with an airport lid, a jar of sterilized LC medium with an airport lid, alcohol in spray bottle, alcohol prep pads, cotton balls, tweezers or knife (optional), permanent marker.

1. Gather all tools and supplies on a clean surface.
2. Spritz hands with alcohol and rub together until dry.
3. Unwrap syringe, but don't uncap yet.

▲ Hand position for preparing to insert needle in SHIP. ISABELLE KIROUAC

▲ Taking biopsy of tissue by stabbing mushroom. ISABELLE KIROUAC

4. Remove foil from jars, spray injection ports of both lids with alcohol and wipe clean with cotton ball.
5. Give a final spritz of alcohol to the SHIP of the water jar. Uncap syringe and quickly penetrate injection port.
6. Tip jar until needle tip is in water, taking care to not wet the filter. Draw up a few milliliters of water.
7. Spritz outside of mushroom with alcohol and wipe with prep pad to sterilize the surface. If very fleshy, use tweezers or knife to start a split and peel away exterior tissue on two opposing sides, without touching tool or fingers to interior tissue. Spritz again.
8. Remove needle from water jar and quickly but carefully stab through the mushroom, avoiding highly textured or spore-bearing tissues (i.e., gills).
9. Look inside the needle to see if you got some tissue inside. If not, try again.
10. If so, give a spritz of alcohol to the LC jar's SHIP, and penetrate with the needle.
11. Push down the plunger, squirting the tissue sample and water into the culture media.

12. Aerate, label with species, strain, date, TC (tissue culture), and medium, and set to incubate.
13. Wait to aerate again until visible growth from tissue sample is observed, then aerate daily until myceliated.

Making a Spore Syringe

Once a spore syringe is transferred to LC, grain, or substrate, the result will be *multispore germination,* which leads to multiple genets growing together. Choice fruits from this lineage can be tissue-cultured to isolate a good strain.

You'll need: a spore print in a bag (see "Spore Prints," above), a sterilized water jar with airport lid, cotton balls, a sterile syringe, alcohol spray, and an alcohol flame or substitute.

1. Gather supplies on a clean work surface and sanitize hands.
2. Unwrap syringe, but don't uncap yet.
3. Remove foil from jar, spritz SHIP with alcohol, and wipe clean with cotton ball.
4. Give a final spritz of alcohol to the SHIP of the water jar. Uncap syringe and quickly penetrate injection port.
5. Tip jar until needle tip is in water, taking care to not wet the filter. Draw up a syringeful of water. Leave needle in port for now.
6. Sanitize outside of spore print bag and remove the needle from the jar. With bag on worksurface, stab bag (careful not to go through both layers) and inject water into bag.
7. Use your hand to mix spores into water, then draw water back into syringe.
8. Flame-sterilize needle and cool by squirting out a few drops of spore solution and recap; or, if you want to make multiple syringes, spritz water jar's SHIP with alcohol and inject spore solution in there.
9. Label and wait 24 hours for the spores to hydrate before using. Spore syringes can stay viable for months to years. Store in a cool, dark place.

▲ Drawing up liquid culture.
ISABELLE KIROUAC

▲ Preparing to transfer liquid culture.
ISABELLE KIROUAC

▲ Transferring the liquid culture.
ISABELLE KIROUAC

LC or Spore Syringe to LC Transfer

You'll need: spore/LC syringe and alcohol flame or extra sterile needle, OR sterile syringe and LC jar; sterile LC media jar; alcohol spray; and cotton balls. If using pre-made spore or LC syringe, skip steps 5 and 6 and do step 7.

1. Gather all tools and supplies on a clean surface.
2. Unwrap syringe, but don't uncap yet.
3. Aerate both jars vigorously for about 30 seconds, breaking up the mycelium in the LC.
4. Remove foil from LC media jar, spray injection ports of both lids with alcohol, and wipe clean with cotton ball.
5. Give a final spritz of alcohol to the SHIP of the LC jar. Uncap syringe and quickly penetrate injection port.
6. Tip jar until needle tip is in LC, taking care to not wet the filter. This can be done in midair or on the worksurface. Draw up 1–10 ml LC, making sure you get some mycelium and not just broth. If the needle clogs, push a little out to clear it and continue. Set the jar upright.
7. If using a pre-made LC or spore syringe, either swap its needle for a sterile one or flame-sterilize needle until red-hot. Allow 3–5 seconds for the needle to cool.
8. Spritz the media jar's SHIP with alcohol and quickly insert the needle into the media jar. Squirt contents of syringe into jar.

▲ Scratching mycelium from agar into water. WILLOUGHBY AREVALO

9. Label with species, strain, date, media, source jar and generation, and set to incubate.

Agar Plate or Slant to LC Transfer

You'll need: an aseptic transfer space (i.e., still air box), a sterile syringe and needle, a sterilized jar of water fitted with an airport lid, a mycelial culture on agar (in either a Petri dish or a culture tube [slant]), a jar of sterilized LC medium, alcohol spray, cotton balls, plate wrapping material, and a permanent marker.

1. Load all supplies into clean SAB or other aseptic transfer space. Mist inside box and allow air to settle.
2. Unwrap syringe, but don't uncap yet.

3. Remove foil from jars, spray injection ports of both lids with alcohol and wipe clean with cotton ball.
4. Give a final spritz of alcohol to the SHIP of the water jar. Uncap syringe and quickly penetrate injection port.
5. Tip jar until needle tip is in water, taking care to not wet the filter. Draw up 1 ml water, leaving needle in SHIP.
6. Unwrap agar plate/slant and open it, taking care not to place hands over contents. For plate, open it like a clamshell or by lifting lid and sliding it back away from hand.
7. Without touching the needle to the edge of Petri dish or culture tube itself, squirt a drop of water onto a choice section of the agar, trying to keep it in a discrete bead. Scratch the mycelium underwater with the needle until it is freed from the agar, then suck the now-myceliated water back into the syringe. Recap agar plate/slant.
8. Give a final spritz to the LC media jar, then penetrate with needle and inject syringe contents.
9. Rewrap the plate/slant with parafilm or plastic film.
10. Aerate and label the new LC and set it to incubate.

Grain or Sawdust Spawn to LC Transfer

This is a way to rescue a lost culture if the LC becomes contaminated, dropped, etc. You could still try using the spawn after this procedure, but do so right away.

You'll need: a jar of pure grain or sawdust spawn with an airport lid, a sterile 60 ml syringe (or smaller, if that's all you have), a sterilized jar of water with an airport lid, a jar of sterilized LC medium, alcohol in spray bottle, alcohol flame or equivalent, cotton balls, and a permanent marker.

1. Gather all tools and supplies on a clean surface.
2. Shake spawn jar to loosen.
3. Unwrap syringe, but don't uncap yet.
4. Remove foil from jars, spray injection ports of all lids with alcohol, and wipe clean with cotton ball.

5. Give a final spritz of alcohol to the SHIP of the water jar. Uncap syringe and quickly penetrate injection port.
6. Tip jar until needle tip is in water, taking care to not wet the filter. Draw up a syringeful of water.
7. Spritz SHIP of spawn jar with alcohol and quickly move needle from water jar into spawn jar. Squirt *almost* all the water in. Withdraw needle and cap it.
8. Without wetting the filter, shake/swirl to suspend mycelium in the water.
9. Uncap and flame the needle until red-hot, then squirt out remaining water to cool.
10. Spritz spawn jar's SHIP with alcohol and reintroduce the needle.
11. Tilt jar until needle tip is in the liquid. Draw up as much of the slurry as possible.
12. Spritz SHIP of LC media jar, then quickly move the needle from the spawn jar to the LC jar. Squirt mycelial slurry into the water and remove needle.
13. Label new LC jar and set to incubate.

LC Aeration and Maintenance

Mycelium is aerobic (oxygen-loving) and doesn't normally grow underwater, but regular aeration of the broth will keep the mycelium happy. One of the advantages of LC is that it allows growth in three dimensions, as opposed to the 2-D surface of agar, giving exponentially more mycelial mass. The agitation of the mycelium, aided by the marble, broken glass, or stir bar in the jar breaks the hyphae, resulting in more hyphal tips from which growth advances. Aeration can be done manually by holding the jar just below the lid and swirling steadily to create a whirlpool effect within. This takes practice, but seasoned wine drinkers will already be proficient. Care must be taken not to slosh and splash LC up onto the air filter, as a wet filter acts as a conduit for contaminants, greatly compromising the purity of the culture. During incubation, aerate for 30 seconds per day. Once mycelium fills most of the jar, growth slows and aeration

▲ Shiitake fruited repeatedly from this jar of coffee-laced LC. WILLOUGHBY AREVALO

▲ Cloudy LC contaminated by bacteria. WILLOUGHBY AREVALO

▲ Overgrown but healthy enoki LC. WILLOUGHBY AREVALO

▲ Chicken of the woods LC with a mat of sporulating mold on top. WILLOUGHBY AREVALO

AAAGH! I GOT MY FILTER WET! WHAT DO I DO?
This happens at least once to most people working with LC. When it happens to you, immediately, or as soon as possible, draw up a syringeful of LC from the jar. Transfer it to a jar of sterile LC media if you have it, or cap, label, and store the syringe in the fridge until you can make some sterile LC media.

can become less frequent. If mycelium becomes starved for oxygen, it will float to the surface and create a mat that can become quite thick and tough if not broken up soon.

Over time, liquid cultures will usually become dense, overgrown, and hard to suck up with a syringe. They may even fruit from the surface of the liquid. To prevent overgrowth, store LC jars in the fridge. Slowing their metabolism this way also reduces their need for oxygen. I have too many LC jars to store in my fridge or agitate more than once a week once grown, so mine tend to get overgrown. I do LC-to-LC transfers periodically to make sure I have viable culture to work with. But even overgrown and stored at room temp, I have had cultures stay viable and pure for up to three years.

Aeration can be automated with the use of Teflon-coated magnetic stir bars (or homemade substitute) and a magnetic stir plate. Used stir plates are available online, or they can be made using easily salvaged computer parts and other common materials (though they can be tricky to dial in to work properly). Multiple stir plates can be wired together for simultaneous use. You can find designs at shroomery.org/68/Do-it-yourself-magnetic-stirrers and elsewhere online.

Recognizing Contaminants in Liquid Culture

One of the biggest drawbacks to using liquid culture media is the challenge of identifying and treating contamination in the LC. Some liquid media ingredients cause cloudiness when agitated, potentially causing confusion. Yeasts and bacteria multiply rapidly, then they are fairly easy to see because they cause the LC to be cloudy at rest. You can try to combat bacterial contamination by adding of a few milliliters of 3% hydrogen peroxide to a hazy jar. Use a sterile syringe and follow sanitation procedures as for injecting LC.

Molds can be trickier to distinguish in LC. Their mycelia are white like mushroom mycelium, but they are often a bit thicker and more inclined to float on the surface of the liquid. Mushroom mycelium will also float to the surface of the LC if the liquid is not adequately aerated, or if the culture is overgrown. If you can't tell if it is mushroom or mold mycelium that is floating on top of your LC, incubate without aerating for several days. If it is mold, colorful spores will probably develop by the third to fifth day. To test the purity of your LC, squirt a drop on agar medium (it can be in a small jar with an airport lid) and observe what grows.

AGAR CULTURE

Agar, also called *agar agar*, a gelling agent derived from seaweed, is mixed with water and nutrients, then sterilized, and poured into Petri dishes (also called *plates*). Once cooled, the medium solidifies. Mycelium or spores are grown on the surface, and this has traditionally been the inoculum for grain spawn since the early 20th century. Agar work must be done in an aseptic transfer space. While LC may be a better everyday inoculum for most home cultivators wanting to make spawn and grow a few mushrooms, agar is a more secure method for production at scale. It has unique uses for strain isolation, strain development, bioassays, and other advanced techniques that are beyond the scope of this book.

Even if LC is your primary inoculum, agar still has its uses. LC itself is tested for contaminants by transferring to agar.

PETRI DISHES AND ALTERNATIVES

Most Petri dishes come in sterile packs of 20 and are made of polystyrene, which is not autoclavable nor reusable. Glass dishes are infinitely reusable (until they break), and autoclavable (so they can be poured before sterilization), but they tend to be pricey. Alternatives include baby food jars, ½ cup or widemouth half-pint jars, small flask-shaped liquor bottles, and various polypropylene containers. An advantage of these alternatives is they can be poured before sterilization. If capped with an airport lid, one can do open-air transfers with a needle and syringe (similar to "Agar Plate or Slant to LC Transfer," above).

▲ Elm oyster mycelium growing on malt dextrose agar. WILLOUGHBY AREVALO

Contaminated agar plates can be cleaned up by subculturing (see below), allowing for the isolation of a pure culture; LC cannot be purified this way. Agar allows for the tissue culturing of tough or thin-fleshed mushrooms that are prohibitively difficult to clone with the needle, or of mushrooms that harbor high levels of contaminants. And the mycelium, molds, and bacteria on a 2D field of agar paint a clear picture from which cultivators can learn a great deal.

Preparation of Agar Medium

You'll need: an aseptic transfer space; media ingredients, a gram scale; a wide-bottomed, narrow-necked glass bottle that pours well and fits in your pressure canner (or a number of them, depending on batch size)—a two-liter growler is good; a wad or two of unabsorbent cotton or polyfill; foil; pressure canner (PC); freshly laundered or alcohol-soaked hand towel; sterile Petri dishes (or substitute); scissors; parafilm or substitute cut to lengths (two squares is perfect, double-wrap for storage to prevent drying); and new zip bags or tape.

1. Measure all ingredients. Divide if using two bottles. Bottles should not be more than 75% full by volume.
2. Add dry ingredients to bottle, then hot water. Swirl until fully dissolved.
3. Stuff bottleneck with cotton and cover tightly with foil.
4. Load into PC and sterilize for 30 minutes at 15 psi (see "Substrate Treatments: Sterilization" in Chapter 7).
5. During sterilization, thoroughly clean and sanitize the transfer space.
6. Allow PC to depressurize slowly. If you need to move the PC, do so when pressure is between 2–3 psi. Open when pressure is at 1 psi, either in flow hood or just beside glove box. If pressure drops below 0 psi, forming a vacuum, soak a paper towel in alcohol and drape it over the petcock (valve) before opening to release vacuum, thereby preventing contamination.
7. With a clean oven mitt, remove bottle(s) and place in transfer space. Load other supplies in as well. If using glove box, disinfect and let air settle.
8. When the agar bottle is cool enough to hold bare-handed but still above body temp, it is time to pour (100°F/38°C is perfect). Clean and sanitize hands/gloves, the outsides of Petri sleeve, and scissors.
9. Cut open the bottom of the sleeve of dishes and stand the stack upright. Carefully remove the sleeve and divide stack into 2–5 even stacks.
10. Remove the foil and cotton plug. With thumb and middle finger of receptive hand, lift the lid of the bottom plate and the whole stack with it, and move aside enough to pour. With the active hand, pour agar in, filling about halfway. Keep working from the bottom up until you do all the plates in each stack. Keeping the plates stacked while cooling reduces condensation.

▲ Pouring a stack of agar plates.
CARMEN ELISABETH

11. Allow agar to cool and solidify in the clean space. Label with date and media (use shorthand).
12. Except for the plates you plan to use right away, wrap perimeter of plates in parafilm, then bag in either the original sleeve sealed well with tape or impulse sealer, or in zip bags.
13. Well-wrapped plates can be stored for a couple of months or more. Don't use if any contamination is visible.

> **SELECT AGAR MEDIA RECIPES**
>
> A 1-liter batch is enough for 30–50 standard-sized plates. Adjust batch size accordingly. Dry ingredients should be added to hot water or heated to dissolve. A huge list of media recipes can be found in the Bibliography. Antibiotics can be added to the media after sterilization and before pouring (when media is less than 110°F/ 43°C) to suppress bacterial growth. This can aid in isolating cultures from dirty or waterlogged wild mushrooms, or in cleaning up contaminated cultures. Antibiotics are not recommended when not necessary. The addition of a small amount of future substrate(s) such as ground grain or sawdust to the media can help the cultures to bounce back from transfers more quickly.
>
> Malt Extract Agar (MEA) or Malt Yeast Agar (MYA)
> 1 liter water
> 20 grams agar
> 20 grams dry barley malt extract
> 2 grams nutritional yeast (for MYA, adds beneficial vitamins and amino acids)
>
> Potato dextrose agar (PDA)
>
> Gently boil 200 grams unpeeled, chopped, preferably organic potatoes in 1 liter water for 30 minutes, then strain through fine filter. Add 20 grams agar and 20 grams dextrose (corn sugar) to the decoction, and heat until dissolved. Dilute with enough water to make a total volume of 1 liter.
>
> Dog Food Agar (DFA)
> 20 grams agar
> 20 grams dry dog food
> 1 liter water

▲ Splitting mushroom to expose clean tissue. MAX KIRCHGASSER

▲ Cutting out a piece of tissue. MAX KIRCHGASSER

▲ Plating the tissue. MAX KIRCHGASSER

▲ Wrapping plate with parafilm. MAX KIRCHGASSER

Tissue Culturing to Agar

You'll need: an aseptic transfer space, a mushroom—use fresh, young specimens when possible (have a few on hand if a tricky size or texture), several blank agar plates (the more attempts you make, the greater your chances for success), alcohol in spray bottle, alcohol prep pads, clean knife, (ideally sterile) tweezers and/or scalpel, alcohol flame or substitute, sterile jar with 3% hydrogen peroxide (optional), parafilm or substitute, and a permanent marker.

1. Load all supplies and tools into transfer space. Sanitize and allow still air to settle or flowing air to blow clear.
2. Sanitize hands or gloves.
3. Unwrap plates if wrapped.
4. Sanitize surface of mushroom with alcohol spray and prep pad.
5. Tear mushroom open to expose interior tissue that has never touched air. If needed, use knife to start tear, but don't slice, as the

▲ Max Brotman and a plate of *Morchella angusticeps* (black morel) culture at the Bay Area Applied Mycology lab in Oakland, California. WILLOUGHBY AREVALO

blade would drag in contaminants. If necessary, set down mushroom so interior tissue doesn't touch anything.

6. Flame-sterilize scalpel or tweezers until red-hot. Open a plate clamshell style, and without touching the tool to the plate, dip the tip in the agar medium to cool. Close lid.
7. Without setting down tool, pick up the mushroom and cut/tweeze out a tiny piece of tissue. Avoid spore-bearing tissues.
8. If specimen is suspected of harboring large numbers of bacteria, tissue can be dunked for a few seconds in 3% hydrogen peroxide. If dried, mushroom can be rehydrated in it for up to a minute.
9. Open the plate, clamshell style. Without touching the tissue or tool to anything but the agar, transfer the tissue sample to the agar, ideally embedding it in the agar. To save on plates, transfer three samples to each (optional).

10. Repeat for each plate. Wrap, label (species, strain, TC, and date), and incubate plates in as clean a space as possible, at around 75°F/24°C.
11. Monitor growth every day, and subculture before the culture reaches the edge of the plate—or right away if contamination is observed.

Subculturing Agar (Agar to Agar Transfer)

Subculturing (or *subbing*) is the act of transferring a small piece of myceliated agar to a new agar plate. Subbing is a way to help mycelium outrun contaminants; it is also a way to turn one healthy plate into a bunch of healthy plates. If doing a multispore germination on agar (see below), subculture from the healthiest-looking sector to isolate a single dikaryotic strain. Always subculture from the leading edge. Subbing a culture repeatedly can lead to senescence (aging) and loss of vigor. Varying the media recipe helps counter this, as does transferring to liquid then back to agar. Always label with the number of transfers the culture has been through, i.e., T1, T2 etc., as well as species, strain, date, and media. Avoid transferring more than 8 times. Keep the source plates as backups unless contaminated.

Inoculating Agar with Spores

This will result in the formation of multiple strains, which can be subbed to isolate a choice strain. You'll need: an aseptic transfer space; alcohol; spore print (see "Spore Prints," above); inoculation loop (can be made out of thin wire attached to a handle); alcohol flame or substitute; blank agar plate(s); parafilm or substitute; and a permanent marker.

1. Load supplies into transfer space. Sanitize surfaces, outside of spore print bag, hands, and air.
2. Flame-sterilize inoculation loop and cool by dipping in agar (open the plate clamshell style).
3. Open spore print bag and scrape up some spores with loop.
4. Clamshell plate again and streak spores across plate with a snaking stroke.

▲ Cutting shaggy parasol stem-butt tissue onto cardboard that was soaked in hot water and drained, with one face removed to expose corrugations. Care is taken to exclude any bugs or larvae. ISABELLE KIROUAC.

▲ Layers are stacked (or rolled like sushi), bagged, and labeled. A cotton ball is zipped in a corner as an air filter (optional). WILLOUGHBY AREVALO.

▲ Two days later, mycelium grows out from tissue. Transfer from leading edge once mycelium reaches plain cardboard. WILLOUGHBY AREVALO

5. Wrap, label, and set to incubate in as clean a space as possible, at around 75°F/24°C. Spore print can be resealed and stored again for later use.
6. Monitor growth every day, and subculture before the culture reaches the edge of the plate—or right away if contamination is observed.

Cardboard Culture

Cardboard culture is a convenient and discreet way to gather and transport cultures, though their window of viability is limited. Pieces of cardboard cultures can be dunked in hydrogen peroxide and transferred to agar for isolation. If this is not possible, consider dividing layers and feeding with more cardboard. Some mushrooms may fruit from cardboard alone. To use as spawn, sandwich between short log rounds, bury in moist sawdust, or mix pieces with prepared straw. Expect slow growth, at least at first. The meager nutrient availability in cardboard leads to low contamination rates, especially when made with care and cold-incubated to inhibit competitors (in fridge or cold room). All you need is: plain, brown corrugated cardboard; a zip bag or other airtight vessel; mushroom stem butts and/or pieces of myceliated substrate (wild spawn); and a cotton ball (optional).

LONG-TERM CULTURE STORAGE METHODS

Agar *slants* are a traditional storage method, wherein culture tubes are filled ⅓ full of mixed agar medium, plugged with cotton, covered with foil, and sterilized, then propped up at an angle to cool so the agar solidifies with maximum surface area. The slants are inoculated with a healthy piece of myceliated agar in aseptic conditions, and the cap is screwed on and wrapped in parafilm. They are incubated for a week at 75°F (24°C) then stored at 35–40°F (2–4°C). Slants should be transferred to plates and checked for viability, then returned to new slants every 6 months. Adding a thin layer of sterilized mineral oil on top of the mycelium can extend storage viability into the years. While these time frames are best practices, cultures might remain viable much longer. My friend Paul Kroeger reports having revived shriveled agar cultures that were 20 years old. Wow!

Small amounts of grain or sawdust can be put in a jar with an extra-large air filter hole (to increase air exchange), inoculated, myceliated, and then allowed to dry out in the jar. Once dehydrated, store in a cool dark place, but not the fridge. Such dehydrated cultures can be kept a year or more. To revive, either transfer a single grain or piece of sawdust to agar, or inject at least enough sterile LC media to cover, then incubate. Once growth is observed, agitate the jar, draw up the LC, and transfer it to a new LC jar.

Saving clean spore prints sealed and refrigerated is an excellent backup, as they can remain viable for many years. Sharing cultures with friends and other cultivators is one of the best ways to ensure that a strain will not be lost.

A long-term storage method for liquid culture is to transfer a few milliliters of LC to a jar of sterilized, distilled water, and refrigerate it—but do not freeze. Grain spawn can also be submerged in sterile water and stored in the fridge. The lack of nutrients, oxygen, and warmth causes the mycelium to go into dormancy, and it may remain viable for years.

Strain Development by Assisted Adaptation

Like us, mycelium can learn new skills and adapt to new conditions and substrates. We can encourage adaptation and developmental growth by offering our cultures a gentle push on the edge of their comfort zone. When introduced to new and unusual substrate ingredients in increasing concentrations in culture media, a mycelium can learn which enzymes and other metabolites it must produce and in what ratios to digest and metabolize a substance. This can allow a cultivator to make best use of available waste streams, or to target a specific contaminant for the purposes of mycoremediation—the degradation of toxins by mycelial action.

THE MOROCCAN OYSTER

When traveling in Morocco in 2013, I purchased some wild phoenix oysters (*Pleurotus pulmonarius*) from a vendor at a market. They were the size of dinner plates, and the stems and cap surfaces were already getting fuzzy with re-vegetating mycelium. I cultured them to cardboard, knowing that I had over a month of travel ahead of me. I sandwiched in more cardboard weekly and kept it in a plastic bag in my backpack. Upon return to the West Coast, I gave a chunk to my friend Ja Schindler, who isolated it on agar. I didn't have access to a lab, so I ran what I had left to pasteurized straw. It took a long time to grow off the cardboard and through the straw, but once it did, it fruited some gorgeous and aromatic mushrooms. The culture Ja isolated has been passed around among cultivators and is currently available from Myco-Uprrhizal (see Resources).

▲ The first fruiting of Moroccan oyster at Fungi for the People in Eugene, Oregon. JA SCHINDLER

MAKING AND USING GRAIN SPAWN

MAKING GRAIN SPAWN

Grain Selection

Many grains are suitable for spawn-making: rye (my usual choice), wheat, millet, sorghum, popcorn, milo, and unhulled (feed-grade) oats are popular. While commonly used in medicinal mushroom products, brown rice tends to be very sticky and is not recommended for beginners. Organic grains are preferable, as conventional grains may have residue of fungicides and/or herbicides (glyphosate is commonly sprayed directly on a grain crop just before harvest to kill it, making it dry more quickly). Some cultivators use only human-grade grains (myself included), while others opt for cheaper animal-grade grains from the feed store. High-quality grain leads to high-quality spawn; broken kernels, dust, and foreign matter can all contribute to contamination.

Grain Preparation

Grains can harbor astronomical numbers of bacterial endospores, making pasteurization worthless and requiring sterilization times to be long—between 1½ to 4 hours, depending on the vessel size, what preparation method is used, and who you ask. Various grain prep methods exist; here are four, in order of least to most labor:

◀ Healthy grain spawn, ready to use.
MAX KIRCHGASSER

▲ Tossing to drive off steam.
WILLOUGHBY AREVALO

▲ Adding gypsum.
WILLOUGHBY AREVALO

▲ Rye cooked just right and ready to be sterilized. WILLOUGHBY AREVALO

1. Load measured amounts of dry grain and water into jars and sterilize in PC. For each quart jar use 1 cup (240 ml) rye grain and ⅔–¾ cup (170–200 ml) water. For half-gallon jars, use 3 cups (600 ml) rye and 1¾ cups (400–460 ml) water.
2. Pour boiling water over grains in jars. Soak 8 hours, drain, then sterilize in PC.
3. Simmer grain in water until al dente. Drain, load into jars, and sterilize in PC.
4. Rinse grains repeatedly until water runs clear. Soak in cool water 6–12 hours, then drain. Simmer in fresh water until al dente, then drain (consider saving liquid as culture medium). Toss or stir to drive off steam, load into jars ¾ full, and sterilize in PC.

I prefer method 4, as grains are cleaned, evenly hydrated, and the soaking encourages the germination and subsequent destruction of bacterial endospores. No matter how you do it, follow these guiding principles:

- Strive for a moisture content of about 50%, keeping in mind that "dry" grain is roughly 10–15% water. Never allow standing water in the bottom of the jar, as wet grains can lead to bacterial wet-spot contamination. When inoculating with LC, err on the dry side.
- Grains expand during soaking, cooking, and sterilization. Spawn jars should not be more than ¾ full, otherwise breaking apart the spawn will be very difficult.
- The addition of gypsum after draining (if draining) helps reduce stickiness, increases moisture capacity, and supplies essential minerals. Measure 1–3 grams per 100 grams of grain, then dust it on the grains with a fine wire-mesh sieve and mix in.
- Cap jars with airport lids and cover with aluminum foil before sterilizing in PC. Sterilize quart jars 1–2 hours, half-gallon jars 2–3 hours, and 5–6 lb. bags 3–4 hours (for info on using bags, see "Inoculating Sawdust with Grain or Sawdust Spawn" and "Containers for Mycelial Growth and Fruiting" in Chapter 7). Allow grains to cool to room temperature, shake to redistribute moisture and prevent clumping, then inoculate as soon as possible.
- Avoid burst grains, as they are sticky and favor bacteria. Grains should be cooked through, but still be firm and chewy. They will cook more during sterilization.

▲ These triticale grains, prepared with method 4, looked good after simmering, but many burst during sterilization. I loaded my PC super-full for this cook, and the jars that were touching the wall of the PC ended up looking like this from the extra conducted heat. Fortunately, the spawn came out okay anyway. WILLOUGHBY AREVALO

Inoculating Grain with Liquid Culture or Spore Syringe

Follow the same basic procedure as outlined for "LC or Spore Syringe to LC Transfer," in Chapter 5. Inoculate each quart jar of grain with 2–10 ml of LC, or more for larger volumes. Multiple grain jars can be inoculated with a 60 ml syringe. Just flame-sterilize the needle until red-hot between each jar. Swivel syringe around during inoculation to broadcast LC all over the grains. You can refill the syringe during a single session if using the same culture; just be sure to flame the needle before going back into the LC jar. Incubate until fully grown, usually 2–4 weeks. Grains inoculated with LC myceliate faster than grains inoculated with agar. Shaking liquid-inoculated grain jars

during growth to distribute mycelium and speed myceliation is optional—I rarely do it.

Inoculating Grain with Agar

If you're growing LC, this won't be necessary for general purpose spawn production, but it is a good skill to have, and some growers (and mushroom species) prefer agar to LC. One plate ⅔ covered in mycelium is enough for 6–8 quart jars. You'll need: an aseptic transfer space, contaminant-free agar plate(s) ⅔–¾ of the way myceliated, sterilized grain jars or bags, alcohol spray, paper towels, alcohol flame or substitute, and sterile scalpel.

1. Load supplies into transfer space, sanitize hands and outside of grain jars or bags, then unwrap plate(s).
2. Flame-sterilize scalpel and cut agar plate into wedges, including leading edge but excluding the very margin of the agar, which may harbor hidden contaminants.
3. Stab a wedge or two and drop into grain jar or bag, taking care not to expose grain or agar to air for longer than necessary. Avoid touching scalpel or agar to lid or mouth of jar or neck of bag.
4. Close jars or seal bags, label, and incubate until fully grown, about 1–3 weeks. Shake jars or bags after mycelium has grown out a ways from agar (around day 4 or 5) to distribute mycelium and speed growth, and shake again a week or so later if it looks like it needs it. It's a fine balance between stimulating it and disturbing it.

◀ Inoculating a batch of grain jars with LC. ISABELLE KIROUAC

▲ Mycelium reaches toward the last few grains in this LC-inoculated jar. THEO ROSENFELD

▲ Cutting a wedge of agar for transfer to fresh agar or to grain. CARMEN ELISABETH

USING GRAIN SPAWN

Once fully myceliated and free of noticeable contaminants, use grain spawn promptly or store in the fridge until you're ready to use. Mycelium knits all the grains together, so they must be broken apart before use. This is done by shaking and/or banging jars on a tire or upholstered furniture (not your body). While it usually takes some force, be as gentle as you can be to prevent avoidable damage to the

Making and Using Grain Spawn • 95

mycelium. If growing grain spawn in filter patch bags (discussed in Chapter 7), simply break it up with your hands before opening the bag. Grain spawn will turn brown from shaking, as the mycelium gets broken up and matted down. The more finely broken up, the more points of inoculation there will be in the substrate, leading to faster myceliation. The same goes for sawdust spawn. The amount of spawn you will use depends on the mass and nutrient levels of your substrate. For instructions on spawning to fruiting substrates, see Chapter 7.

Grain-to-Grain (G2G) Transfers

Cultivators inoculating their grain spawn with agar tend to use that grain (the grain masters) to inoculate more grain jars, greatly extending their agar cultures. Inoculate grain to grain at a rate of 1:10 in an aseptic transfer space following the procedure as outlined for "Inoculating Sawdust with Grain or Sawdust Spawn," in Chapter 7. Use only the best-looking, contaminant-free grain as masters. Shake 'n break potential grain masters and incubate 24 hours before using. Jars that don't bounce back probably harbor some unseen contaminants and shouldn't be used as inoculum. If using LC, its abundance and the ease of inoculating with it makes G2G transfers unnecessary.

Stick Spawn

Some cultivators, especially those facing food insecurity, use small amounts of grain spawn or LC to create stick spawn, which is done by poking into vessels of fruiting substrate. Soak then drain bamboo skewers or coffee stir sticks in a jar, leaving around 40 ml water in the bottom. Sterilize under pressure for 45 minutes, then inoculate with a teaspoon of grain spawn in an aseptic transfer space. Incubate jars on a slant so the spawn is touching the length of the sticks. Shake after partial growth to distribute mycelium if uneven. Shake before use.

GRAIN SPAWN FOR FOOD AND MEDICINE

Fresh, pure, myceliated grain is edible, and it is the main ingredient in many medicinal mushroom supplements. Once you've learned the skills of making grain spawn, it's an easy and relatively quick way to put lots more fungal biomass into your diet. If you're overwhelmed by the fruiting process, or if your space is very limited, myceliated grain could be a suitable and satisfying end product for you.

A largely unexplored and recently developed food product, myceliated grain is flavorful, protein-rich, and it contains many of the same medicinal compounds and nutrients as fruiting bodies. When grown in tapered 8, 16, or 24 oz jars, a whole block can be slid out, sliced, and used like tempeh (which itself is cooked soybeans inoculated with *Rhizopus oligosporus* mold). Myceliated grain can be substituted for grain, mushrooms, or ground meat in many recipes. Dehydrated, it can be ground into a flour, cooked whole as a grain, rolled and cooked as porridge, or cracked and incorporated into beer recipes. It should be cooked before eating. Certain species taste better than others, so experiment and see what you fancy. I really like king oyster spawn. Different grains and blends can be used, but resist the temptation to eat myceliated beans—some edible mushrooms such as oysters produce toxic metabolites when grown on legumes.

Next time you're at a store selling medicinal mushroom capsules, look at the ingredient list on the bottles—many include myceliated brown rice instead of or in addition to fruiting bodies. Then look at the price tag and think about how little it costs to make a jar of grain spawn. For mushrooms that don't readily fruit in culture, such as many of the medicinal polypores, myceliated grain is the best way to get the active compounds. It can be tinctured and/or decocted (see "Basic Medicinal Preparations: Tincture, Decoction, and Double Extract," in Chapter 9), or thoroughly dried, ground, and encapsulated.

▲ Sautéing oyster grain in place of ground meat for a tomato-based pasta sauce. WILLOUGHBY AREVALO

FRUITING SUBSTRATE FORMULATION AND PREPARATION

CONTAINERS FOR MYCELIAL GROWTH AND FRUITING

Jars

Glass jars in various sizes and shapes are used for LC, grain spawn, nutrified sawdust, and other fruiting substrates. Any jar with a tight-fitting lid can work, but Mason jars are ideal because of the universality of the lids. I usually prefer regular-mouth pints for LC, and quarts or half-gallons for grain and sawdust. Plain jars allow better visibility of the contents than decorated jars. I don't differentiate between my jars for mushrooms and for canning and food storage. Wash thoroughly with hot water and soap or in a dishwasher. Pre-sterilization is unnecessary.

Plastic Bags

Gusseted polypropylene (PP) filter patch bags (called *filter patch bags*) are manufactured specifically for mushroom cultivation with an integrated micron filter patch. Polypropylene can withstand the heat of sterilization, making these ideal for grain spawn and nutrified sawdust. They are a bit pricey and only sometimes reusable. Oven (turkey) bags are made of heat-resistant nylon and can also be sterilized, but they are thin (prone to puncturing) and need to have an air filter added at closure (see below). Seal PP bags well with an impulse sealer (heat sealer), zip ties, or stiff wire. Though typically inoculated with grain spawn and sealed in an aseptic transfer space, you can add a SHIP to PP bags for open-air inoculation with LC by adding a blob of RTV silicone reinforced with clear packing tape and sealing before sterilization. See Resources for sources.

Most clear-plastic bags, like produce bags from the grocery store, are made of polyethylene (PE, poly). PE bags can withstand pasteurization but not sterilization, making them useful for steam pasteurization of sawdust or other prehydrated substrates. I usually reuse washed produce bags for this, but gusseted PE bags are commonly used. A makeshift filter is added to the top of the bag after inoculation (so make sure to leave room). The extra bag is fed through a collar (plastic is used at farms but ⅓ of a toilet paper tube works great), folded over the outside, stuffed with cotton or polyfill, and fixed with a piece of newspaper and a rubber band. A piece of Tyvek can also be used as a filter. Depending on the fruiting preferences of the species and cultivator, the cotton plug is removed after spawn run and mushrooms fruit out the collar, or the collar is removed and the bag is either trimmed to the top of the substrate (for top-fruiters), rolled down and punctured (good for oyster or lion's mane), or removed completely (shiitake).

PE bags are commonly used for growing oysters on straw. Clear or black poly tubing can be purchased in rolls (see Resources), cut to desired length, and tied off at one end to make huge bags (4 mil, 16" is ideal for big "strawsages" [aka straw logs]). Bags are stuffed very firmly with straw and spawn and tied off at the other end, adding a loop of rope or metal ring for hanging. Quarter-inch "X" or "Y" shaped holes are poked every few inches with a sanitized razor (stainless steel, 3-blade archery broadheads are ideal), allowing for gas exchange without excess drying, and, eventually, for fruiting sites. No filtration is needed because oysters are usually faster than competitors on straw.

▲ A polypropylene filter patch bag with added self-healing injection port (SHIP).
WILLOUGHBY AREVALO

Fruiting Substrate Formulation and Preparation • 99

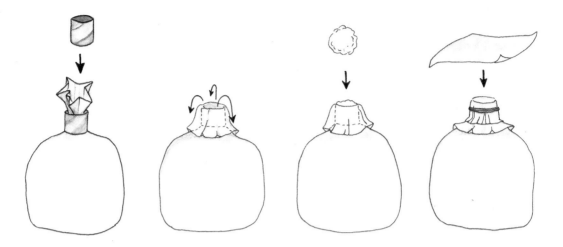

▲ Making a filter for a PE bag: feed bag through collar, fold down bag, plug with cotton or polyfill, cover with paper and rubber band. CARMEN ELISABETH

▶ A magnified view of oyster primordia pushing out through a hole in a PE bag. A ¼" hole can release a massive cluster of oysters. Note straw for scale. WILLOUGHBY AREVALO

Buckets and Other Large Plastic Containers

An alternative to oyster strawsages, 3- to 5- gallon food-grade buckets are stackable, hangable, reusable, recyclable, and easily salvaged from the waste stream. Make arrangements with ice cream parlors, donut shops, or restaurants to get their empty buckets, or just go dumpster diving. Numerous ¼"–⅜" holes should be drilled on the sides for gas exchange and fruiting. Wash and sanitize in 10% bleach solution and air dry. Stuff very tightly with straw or other substrates to prevent gaps and lost fruitings after substrate eventually shrinks (unless you're using coffee grounds, which are easily over-compacted). Buckets, bins, etc. can also be used for other species. Small buckets with a 1" hole in the lid filtered with micropore tape (and no other holes) can be used for shiitake grown on pasteurized sawdust and inoculated with grain spawn. Simply slide block out of the bucket before fruiting.

▲ Wine caps fruiting from alder chips, Douglas fir sawdust, and fermented straw in a nursery pot at DIY Fungi in Victoria, BC. These are past their prime for harvest. DANIELLE STEVENSON

▲ Reishi fruits from a laundry basket at Mush Luv. Nina and Charlie boiled woodchips for 12 hours, drained and cooled them, then inoculated them with 5 lbs. sawdust spawn and 2 lbs. of grain spawn. It incubated three weeks indoors in a ventilated plastic bag, then it was moved outside and watered for fruiting. JA SCHINDLER

▲ (L to R) Bea Edelstein, Klara Russell Kirchgasser, me, and Max Kirchgasser fill expired surgical gloves with pasteurized sawdust and oyster grain spawn for the *Mycelial Connections* sculpture project. CARMEN ROSEN

Trays

Trays are useful for growing top-fruiting species on pasteurized substrates, especially compost or compost-like mixes. Fill with hydrated substrate, cover with foil, steam pasteurize, inoculate, recover, and set to incubate. Apply casing once myceliated (if using) and remove cover when setting to fruit. Many things will work, like baking pans, dish tubs, litter boxes, nursery flats, or trays made from untreated conifer wood.

Other Vessel Options

One can fill many types of vessels with pasteurized substrate. Nursery pots or laundry baskets can be filled with spawn and pasteurized substrate and stacked; just protect them from drying out during spawn run. Unstack pots for top-fruiters and treat like trays, or keep stacked for side-fruiters and slit sides of pots for fruiting sites. Perforated PVC drainage pipes, culverts with holes drilled, or similar tubes can be used as a strawsage alternative. Make sure to pack tightly and cap with something. Biodegradable vessels like burlap sacks or cardboard boxes and cylinders can also be used. They will be eaten by the mycelium as well, which may detract from yields and cause them to fall apart. Other possibilities include woven plastic feed bags, onion sacs, gloves, perforated metal boxes, gourds, and Styrofoam coolers. Whatever you salvage, clean well and sanitize before and between uses.

Air Filter Options

A number of materials are microporous and allow gas exchange without letting in contaminants. Filters must stay dry, otherwise the water acts as a conduit for microbes. Polyfill, which is found in pillows, puffy coats, and stuffed toys, is great because it is cheap (or free) and versatile, and its bulk offers moderate resilience to wetness. Only reuse if clean. Cotton balls or batting works similarly and is biodegradable but is more absorbent and may be grown through by mycelium over time, blocking airflow and increasing vulnerability to contamination. I use it sometimes as filters for PE bags and sawdust jars, but never for LC. Tyvek is thin synthetic fabric that is microporous and fairly water-resistant; it is available as housewrap, painter's suits, protective sleeves, and mailing envelopes. A double layer of micropore medical tape wins for its convenience factor, but it is thin and not at all water-resistant, therefore not suitable for LC jars. Its rayon fibers deteriorate slightly from sterilization in a PC, so replace tape after each use (i.e., on jar lids). Filter discs fitted to Mason jars are manufactured for mushroom cultivation and available from specialty suppliers, but they are expensive compared to the other options.

> **AVOIDING MAKING EXCESSIVE TRASH**
>
> Besides energy consumption, the biggest negative environmental impact of mushroom cultivation is that it usually produces a significant amount of garbage, particularly plastic grow bags. Polyethylene (PE) and polypropylene (PP) bags are recyclable in many places, but they must be cleaned first, which is a pain in the butt and risky for cultivators, who do not want to be washing bags that have harbored contaminants. Used PP bags that weren't contaminated or cut open can be washed, dried, and reused, but they lose integrity after repeated sterilizations (one friend reports five uses with good results). However, they can be reused right away without washing for packing spawned straw and other bulk substrates.
>
> Using alternatives to new plastic is one of the best approaches. Glass jars, plastics borrowed from the waste stream (like salvaged buckets), and other upcycled materials can reduce trash production.

THE SUBSTRATES

Sawdust

For most wood-loving mushrooms, sawdust is the best possible fruiting substrate, especially when supplemented. Most types of hardwood sawdust work well, especially fast-growing and fast-rotting types like alder, cottonwood, poplar, aspen, and birch. Oak and other dense hardwood sawdusts may be a little slower to myceliate but also support excellent fruitings. Other good woods include maple, beech, elm, sweetgum, and others. Most mushrooms dislike conifer wood due to its low-nitrogen and high-turpentine, tannin, and resin content, but it can be used for some species when leached and supplemented, mixed with hardwood, or composted. Douglas fir works better than other conifers. Cedar, redwood, and other members of the *Cupressaceae* (Cypress family) are very antifungal and not to be used at all. Walnut, locust, and most fruit woods are also to be avoided.

Sourcing sawdust can be a puzzle. The best place to get it is from a sawmill, as that sawdust is usually quite pure and of a consistently good particle size—big enough to have some structure but small enough to have good density. Woodworkers are another good source, but the texture is variable, leading to variable results. Sander dust is too fine and can become anaerobic (I recently met a farmer who had just lost 3,600 blocks to contamination due to this). Planer shavings are too fluffy. I've found that a mix of shavings, sander dust, and dust from other woodshop tools is usually adequate for fruiting, but it's hard to work with as spawn for log inoculation. Sawdust from chainsaws is not recommended as it is contaminated with bar oil, a petroleum-based lubricant for the saw. Also avoid any sawdust that contains adhesives, laminates, paint, or other pollutants.

One readily available sawdust source for home cultivators is hardwood fuel pellets, available at hardware and home improvement stores. Pellets take about 30 minutes to expand and hydrate, and

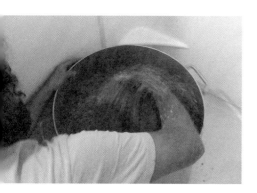

▲ I prefer to mix sawdust by hand while misting with water until perfectly moistened. Here, I use the shower. CARMEN ROSEN

they may require a little kneading. Different brands absorb different amounts of water. One tek calls for 6 cups of pellets and 7 cups of boiling water to be mixed in a filter patch bag, tied off with string, insulated, and left overnight to pasteurize. It is inoculated the next day (open air, ok) with 2 cups of grain spawn. Another tek calls for mixing 5 cups pellets, ½ cup wheat bran and 6 cups of cool water in a filter patch bag, then sterilizing 120–150 minutes at 15 psi in a pressure canner. Inoculate in aseptic transfer space with 2 Tbsp grain spawn.

Plain sawdust can be steam pasteurized; when nutrified (supplemented) with a nitrogen source, it must be sterilized. Either way, it must be prehydrated to field capacity (see sidebar) or just under. Sawdust can be either soaked (<12 hours) and thoroughly drained or have water added to it during mixing until it is just right. The correct moisture content is between 60–65% by weight (when combined with kiln-dried sawdust); supplemented sawdust can normally hold more water than plain sawdust. Dry sawdust is hydrophobic and takes longer to hydrate than aged/damp sawdust. Many cultivators prefer sawdust that has been weathered but not exposed to soil, as it holds more water. Rest 15 minutes after hydrating and check moisture content before filling vessels. If your substrate is too wet, add a little more dry to the mix. Properly hydrated sawdust will stick to hands and form a clump when squeezed.

The addition of 1–5% gypsum improves structure and adds beneficial minerals, boosting the speed and mass of mycelial growth and subsequent fruitings. The addition of 30% wood chips improves yields on second and third flushes but should not be done when making sawdust spawn for log inoculation. Chips should be soaked for 12 hours, then thoroughly drained prior to mixing. Wood chips can be gleaned from arborists and city yards or purchased at hardware stores as smoker chips. Avoid huge chips, those with large amounts of leaves, or those that are composting. Various online services exist to help people get free chips from local arborists.

> *Field capacity* is defined as the most water a substrate can hold against gravity, i.e., fully drained. This is the optimal hydration for most substrates. Water content is different than water availability—water covering the substrate particle surfaces is much more available than water locked up in the cellular structure of the substrate. To check field capacity, firmly squeeze a handful of substrate—it should release just a couple of drops of water, not a stream. Avoid having standing water in the bottom of substrate vessels, especially if inoculating with LC.
>
>
>
> ▲ Overhydrated substrate particles (L); substrate particles at field capacity (R). CARMEN ELISABETH

NUTRIFIED SAWDUST FRUITING BLOCKS

The following is a classic, high-yielding recipe for nutrified sawdust fruiting blocks, adapted for home-scale growers from Paul Stamets's *Growing Gourmet and Medicinal Mushrooms* and Marc R. Keith's *Let's Grow Mushrooms* video series. Here, one bag of moist substrate is equivalent to roughly 3.5 liters, or 2.5 kg (5.5 lbs.), which is the right amount to use for most common PP filter patch bags. Measurements are approximate and may fluctuate depending on particle size and moisture and nitrogen content of your ingredients. Adjust as needed. Measuring by volume is less precise but usually more practical than by weight.

By ambient dry weight per bag:
(total per bag = 1 kg or 2.44 lbs. dry materials)
510 grams sawdust (20 oz.)
260 grams wood chips (10 oz.)
  (or omit and use a total of 770 grams [30 oz] sawdust)
200 grams bran (8 oz.)
30 grams gypsum (.1 oz.)

By volume, measured before or after hydration:
(1 part = 1 cup per bag)
8 parts sawdust
4 parts wood chips
1 part bran
⅛ part gypsum

1. Measure and thoroughly mix ingredients. Either hydrate then mix, or mix then hydrate, to field capacity. If too wet, add a little dry sawdust.
2. Load into PP filter patch bags or jars with airport lids. A few prepared grains can be added to the top of vessels to be inoculated with LC. Bags should be filled halfway or less (5 lbs. for large bags). Jars should be filled 1" (2 cm) below top, or halfway for fruiting in the jar (as for antler-form reishi). Using a hollow cylinder (coffee can with bottom cut out) that fits in the bag or a jar funnel helps keep the tops clean and eases filling.
3. Wipe clean the tops of the bags/jars as needed.
4. Respecting the folds of the gussets, flatten the top of the bag. Insert a Tyvek sleeve to filter the air that will be sucked in after sterilization creates a vacuum within, and fold the bag and sleeve accordion-style.
5. Load bags or jars into PC or super pasteurizer (see "Substrate Treatments: Sterilization," below), putting jar rings or other spacers between and around bags to allow steam flow and prevent bags from touching wall of PC. Place a heavy ceramic plate on top of bags to prevent them from expanding and blocking steam vents/valves.
6. Sterilize in PC at 15 psi for 2 hours, or in super pasteurizer for 10–14 hours at 0 psi, or 8 hours at 2 psi.

▲ Mycelium racing down through alder sawdust supplemented with dried nettle leaf and gypsum and inoculated with liquid culture. WILLOUGHBY AREVALO

Inoculating Nutrified Sawdust with Liquid Culture

Prepare sawdust as usual, load into jars with airport lids or filter patch bags modified with a blob of silicone, and optionally add 1–2 Tbsp precooked grain just below the inoculation port to act as grain spawn, boosting the nutrients available. Bags should be sealed prior to sterilizing, leaving a small amount of air space in the bag to act as a plenum during incubation, but not so much that the bags explode during sterilization, allowing for open-air inoculation (see "Inoculating Sawdust with Grain or Sawdust Spawn," below, for sealing instructions). Follow the same basic procedure as outlined for "LC or Spore Syringe to LC Transfer," in Chapter 5. Inoculate each quart

Fruiting Substrate Formulation and Preparation • 107

jar of sawdust with 10 ml of LC, half-gallon jars with 20 ml and 5 lbs. filter patch bags with 40–60 ml. Aim the stream of LC at the grains, if using, otherwise broadcast the LC across the surface of the sawdust. Label and set to incubate. Remove lid for fruiting, and keep jar upright for top-fruiters or on its side for side-fruiters.

Inoculating Sawdust with Grain or Sawdust Spawn

You will need an aseptic transfer space, sterilized sawdust bags or jars to be inoculated, enough spawn for the job, alcohol spray/wipes, scissors or a blade to open spawn bag (if using), stiff wire or zip ties or an impulse sealer (for sawdust bags), and a marker for labeling. Gloves are optional. As nutrified substrate is very susceptible to contamination, work with a great level of focus, efficiency, and deliberation. After several times doing this, it will become easy.

1. Start by breaking up the spawn. Jars can be shaken or bonked on a firm, soft surface like a rubber tire or upholstered chair. Break up bags of spawn manually before opening. Though it usually takes some force, be as gentle as possible while breaking the spawn into nearly individual particles. When well broken up, each grain is a discrete point of inoculation which will grow outward in three dimensions.

2. Load all supplies into glove box or other transfer space. Move the substrate in as directly as possible, opening the PC in the space to unload if space permits.

3. Thoroughly clean your hands and the outside of the spawn vessel with alcohol.

4. Line up vessels to be inoculated. Loosen jar lids, if using, leaving lids in place. If using bags, unfold bags, remove Tyvek sleeves, and open bags—taking care not to touch the inside of the bags or move your hands over an open vessel. Depending on the height of your glove box or flow hood, you may have to open substrate bags to the side.

5. Open the spawn vessel (if cutting spawn bag, sanitize blade first) and gently pour an equal amount of spawn into each substrate vessel.

Two tablespoons (30 ml) of grain spawn is enough to inoculate a 5.5 lb. (2.5 kg) bag of nutrified sawdust, but more can be used.

6. Close up the inoculated vessels, sealing bags with wire, zip ties, or an impulse sealer, leaving some air space inside for the growing mycelium. If using a flow hood, fill the bags with filtered air before sealing. If using wire or zip ties, fold the bag down twice before rolling in from the corners and tying closed. Ensure that the seal is good by squeezing the bag and listening for the hissing of a leak. Reseal if imperfect.
7. The following steps may be done in the open air: label each vessel, shake to distribute spawn through substrate, form bags into a block shape, and move to incubation space.
8. Leave space around nutrified sawdust bags; if crowded together, they may develop dead zones due to overheating. Unsupplemented sawdust bags can be stacked or crowded because they produce less heat.

Straw

Cereal straws make an excellent substrate for oysters and some other mushrooms. Wheat is used most in North America and rice in Asia, though barley, oat, rye, and others are good, too. Hay is not suitable because the nitrogen level and spore loads are too high, leading to contamination. Straw should be fresh, golden in color, and stored dry. Organic straw is preferred, as conventional straw is likely coated in the herbicide glyphosate (Roundup). Buy from feed or garden stores or direct from farmers.

Straw should be shredded to 1–3" particles before use to increase substrate density and structure, making it easier for the mycelium to get a "bite" of it. If you don't have a chipper/shredder, you can shred with a lawnmower on a smooth surface or a weedwhacker in a trash can. Wear a dust mask and eye protection. Shredding with hand tools is a tedious pain in the wrist, and very sharp blades are needed. When shredding is not convenient, straw can be coiled into fruiting containers to improve structure. Yields can be improved with light supplementation (around 5%), but contamination risk is

> **INOCULATING PASTEURIZED SAWDUST**
>
> Pasteurized plain sawdust is inoculated the same way, though it can be done without an aseptic transfer space, and inoculation rates should be higher to compensate for lower nitrogen levels in the substrate, about 1–2 cups (250–500 ml) of spawn per 5.5 lbs. (2.5 kg) substrate, if you want a good yield. Less spawn can be used if the substrate will be used for spawn rather than fruiting. Because polyethylene bags are usually used for pasteurized sawdust, they will be closed with a collar and filter rather than being sealed. See "Containers for Mycelial Growth and Fruiting," at the beginning of this chapter.
>
>
>
> ▲ Inoculating pasteurized sawdust with grain spawn. MAX KIRCHGASSER

also increased. Straw is best pasteurized in hot water, or by other bath-type methods; see "Pasteurization and Alternatives," near the end of this chapter.

Other Agricultural Wastes, Weeds, and Invasives

Certain mushroom species, such as oysters, wine caps, and others, are very flexible and will eat many plant materials, offering a great opportunity for creativity. However, some mushroom-substrate combinations will support only mycelial growth but not fruiting. Here is an incomplete list of potential substrates that have been successfully used alone or in combination with other materials: cottonseed hulls, peanut shells and plants, sunflower stalks, cannabis/hemp stalks and roots, corn stalks, cobs and husks, water hyacinth, nettle stalks, English ivy vines, Japanese knotweed stalks, kudzu, European beach grass, Scotch broom, garlic stalks, squash plants, dried fruit peels, bean pods and straw, hedge laurel, blackberry and raspberry stalks, hops plants, grape vines, grape pomace, banana leaves, cacti, coconut coir, burdock stalks and burrs, lawn clippings, tree and shrub leaves and branches.

▲ Shredding corn stalks in a barrel with a weedwhacker. KAITLIN BRYSON

Anything leafy and green should usually be dried before processing, while woody stuff can be used fresh. Most of these types of substrates should be shredded like straw then treated by pasteurization or an alternative. Many agricultural by-products are useful as supplements to boost nitrogen in substrates. Soybean hulls, cottonseed meal, various brans, and alfalfa meal are all commonly used, but there are hundreds of other possibilities.

For instructions on inoculating straw, ag wastes, etc., see end of this chapter.

Paper Products, Coffee Grounds, and Other Urban Waste Streams

When we open our minds and eyes to plant-based materials, we begin to see potential substrates all around us. Cities offer abundant streams of substrates for growing and supplements for boosting

▲ Pasteurizing samples of branches of vines and sticks of various plants, including invasive species, to test their suitability as both oyster mushroom substrates and myceliated weaving materials. WILLOUGHBY AREVALO

nitrogen and nutrients in other wastes. Paper products, such as cardboard, egg cartons, shredded office paper, newspaper, books, and (unused) paper or sawdust-based kitty litter, are relatively low in nutrients other than cellulose. Therefore, they can be used with little to no treatment, but yields are improved by supplementation and treatment. Natural fiber rope and twine, burlap, cotton (even old clothes), and many other fiber and textile wastes are suitable, and they lend themselves to creative applications of mycelium in art,

> The pH (relative acidity) of fruiting substrates is important. Though each mushroom tolerates a range that encompasses all the normal substrates, we may encounter issues when we play around with odd substrates. Very acidic substrates at preparation may indicate microbial activity. Most cultivated mushrooms like a slightly acidic pH and slightly acidify the substrate further as they grow. I never test the pH of my substrates, but if you're inclined, pH test strips are available at pharmacies and garden stores.

bioengineering, and more. Some of the many useful nitrogen supplements commonly discarded from industry, food service, and maybe even your own kitchen, include brewery and distillery waste, tea leaves, okara (tofu dregs), and juicer pulp.

Freshly used coffee grounds can be sourced in any city or town any day of the week and are an excellent substrate for acid-tolerant oysters and other mushrooms. Their acidity can be buffered with the addition of about 2% hydrated lime, rendering them suitable for other species such as reishi and shiitake. A "perk" of used grounds is that through the brewing process, they are essentially quick-pasteurized and need no further treatment if inoculated promptly; they can also be inoculated in the open air. Coffee grounds tend to be small particles and are sometimes already wetter than field capacity, therefore they are prone to overcompaction, overhydration, and anaerobic conditions, favoring bacteria and molds. The addition of dry wood shavings (the fluffy kind) and/or corrugated cardboard help to absorb excess moisture and increase aeration. Using small containers can also help to prevent these problems. Gypsum is also helpful in maintaining good structure and for its beneficial minerals.

Since no mushrooms have independently evolved to live on coffee grounds, a little assistance may be required to get a strain to be happy living on them. By adding small amounts of coffee grounds to culture media and spawn, the cultivator can aid this adaptation process. Fortunately, caffeine is a mutagen, increasing the dynamism of the genome. Once a strain is thriving on coffee, make a tissue culture from a mushroom that fruited from coffee, and expand it on coffee-laced or coffee-based media. A similar process can be done for other odd substrates as well.

Compost and Compost-like Mixes

Button mushrooms and portobellos (*Agaricus bisporus*) are conventionally farmed on a thermophilic (hot) composted blend of straw, manure, gypsum, and other additives. Stamets's cultivation books have thorough instructions for making high-quality composted

▲ Blue oysters fruiting from coffee grounds at Wildwood Ecology Labs in Powell River, BC. THEO ROSENFELD

substrate, as do many older cultivation manuals. *The Ground Rules*, by Klehm and Blecher is one of many good resources for thermophilic composting info (see Bibliography).

During the hot composting process, the internal temperature of a pile should climb as high as 140°F (60°C), then drop. As soon it has dropped to 80°F (27°C), it is ready for spawning. If you aren't prepared to use it all right away, or if the max temp didn't get high enough, it should be hydrated and steam pasteurized before spawning. Municipal compost can be used, but it should be steamed first because it may contain pathogens. Bagged compost may also work, but expect mixed results.

▲ A cobra-formed oyster digests and fruits from junk mail and household paper waste. MAX BROTMAN

THE JUNK MAIL DIGESTER

When I was living in a really dirty artist loft in Arcata, I had an old, dried-out bag of oyster sawdust spawn. It had never been opened, and eventually it got moldy on the upper surface. Determined not to waste it, I tore off the moldy surface, soaked my stash of used brown paper bags in the kitchen sink, tore them up with my unwashed hands, mixed in the crumbled-up spawn, and stuffed it all into a used plastic bag. In a week, it was fully myceliated with no sign of contamination, and in two weeks it fruited. I gave the bag to my friends Max Brotman and Claire Brown, who were living in a really dirty punk house in Oakland. Their many roommates past and present got piles of junk mail, so they wet the pile down, tore it all up, and mixed it with the contents of the bag in a big plastic bin. The mycelium took to it well, and more junk mail and other paper waste was thrown in regularly, none of it pasteurized. Mushrooms fruited occasionally from the rubbish, and after about 7 months, it was finally retired to the worm bin en route to the garden beds.

This example illustrates two points about the resiliency of mycelium. First, that its immune system can learn to tough it out in a diverse and potentially antagonistic microbial landscape when introduced to it gradually. Second, that mycelium can transform garbage into food for us, which though not the most appetizing, may be a blessing in dire situations.

Compost-lovers such as Agaricus, shaggy manes, blewits, and others can also be grown on a variety of alternative compost-like blends. Hydrate ingredients to field capacity and steam pasteurize separately, then mix together at spawning.

The Casing Layer

A *casing* layer is a moisture-holding, microbe-rich soil-like covering that is added to the top of some substrates once they are fully myceliated. Most compost-lovers and some other mushrooms require it for good primordia formation, including Agaricus, blewits, shaggy manes, and wine caps. A few wood-lovers like king oysters and pioppino may benefit from, but do not require, a casing.

Common casing ingredients include peat, vermiculite, coir (buffered with 5% hydrated lime or chalk), and gypsum, but many of these are overmined and not renewable. Pea-sized particles of *biochar* (see sidebar) make an awesome casing and is infinitely reusable. Some cultivators use nutrient-poor potting soil. Old cultivation manuals recommend using sandy or loamy soil. Casings should be wetted until a squeezed handful yields a brief stream of water. As casings often harbor Trichoderma and other contaminants, low-temp pasteurization is recommended. Pack moist casing into bags or jars and steam or water bath for 1 hour at 130-145°F (54-63°C) to kill molds but preserve the bacteria that promote fruiting.

Once prepared, spread the casing over the upper surface of the myceliated substrate to a depth of about 2-3 cm (½" to 1½"), making a level but rough surface. Cover with perforated plastic wrap and re-incubate for a week or two, then uncover and move into the fruiting space as soon as mycelium covers the casing. If molds develop, increase FAE and mist lightly with 3% H_2O_2. Maintain very high humidity (90-95% RH) until primordia have formed, then reduce or stop misting as mushrooms develop. Post-harvest, refill any holes or gaps in the casing and try for a second flush.

SIMPLE FORMULAS FOR COMPOST-LIKE BLENDS (BY VOLUME)

60% dry horse manure
25% coconut coir
10% vermiculite
5% coffee grounds

90% shredded straw
10% worm castings

60-70% shredded straw
30-40% leached horse or cow manure

Biochar is a carbon-sequestering soil amendment invented by ancient Amazonians; it is easily made at home by *pyrolyzing*—burning plant matter with restricted oxygen—resulting in a stable, porous material that holds water, air, and beneficial microbes. Weeds, wood, manure, ag wastes, and spent mushroom substrates all make excellent biochar.

SUBSTRATE TREATMENTS

Sterilization

Sterilization, the eradication of all life—including dormant endospores—is achieved through sustained heat at or above 250°F (121°C). Because this is hotter than boiling water and because dry heat is not recommended, this is usually achieved in a pressurized steam chamber such as a *pressure canner* (PC).

Pressure canners are basically big heavy-duty pressure cookers capable of holding greater pressure. The pressure can be accurately regulated—some PCs regulate the pressure with a toggle valve called a *petcock* or *stopcock,* coupled with a pressure gauge. Others have a nipple that is fitted with a weight that will allow the release of the right amount of steam to maintain a consistent internal pressure, as long as the steam is lifting the weight and causing it to rock or jiggle, giving a steady knocking sound. Both petcock types and rocker types usually also have a safety valve, which will automatically open and blow off some steam if the pressure becomes dangerously high. Either type of PC is suitable for mushroom cultivation, though petcock types are choice. Petcocks, pressure gauges, replacement gaskets, and other parts can be purchased from specialty suppliers. Most PCs are for stovetop use, compatible with normal home ranges. Some high-end models contain an internal electric heat source and are programmable.

To achieve sterilization temperature, the PC must be pressurized to 15 psi at sea level. At high altitude, greater pressure is needed to compensate for the lower ambient pressure. See table for the recommended pressure for your elevation. Pressure of rockers can be boosted by adding washers or small magnets to the weight. Use caution. Liquids take a minimum of 15 minutes to sterilize, but, because they contain air pockets, solid substrates must be cooked longer to ensure that the core is adequately heated.

Because of the high temperature and pressure of a PC, it is potentially dangerous, and using one demands care and great attention. However, when treated right, a pressure canner can work safely and

▲ My trusty and huge old pressure canner. It can hold 16 quart jars. I scored it at a flea market for $35, about a tenth of what it's worth. WILLOUGHBY AREVALO

effectively for decades. A nice old PC in good condition may be better than a cheapo new one—they don't make 'em like they used to.

Sterilization at Altitude

ALTITUTE (FT)	ALTITUTE (M)	AIR PRESSURE (PSI)	IDEAL GAUGE PRESSURE (PSI)
0	0	14.7	15
1,000	300	14.2	15.3
2,000	600	13.7	16.0
3,000	900	13.2	16.3
4,000	1,200	12.7	17.0
5,000	1,500	12.2	17.3
6,000	1,800	11.8	17.9
7,000	2,100	11.3	18.4
8,000	2,400	10.9	18.8
9,000	2,700	10.5	19.2
10,000	3,000	10.1	19.6
20,000	6,000	6.75	22.95

COURTESY OF SHROOMERY USER TELADI

Super Pasteurization

This method is basically a long, hot steam bath at (nearly) boiling temperature 205–212°F (96–100°C). Consistent heat of the substrate core for 10–14 hours is sufficient to achieve sterilization. This is a good option for sterilizing big batches of nutrified sawdust. Super pasteurization can be done inside an insulated chamber injected with steam from an external source or in a food-grade 55-gallon drum on a propane burner. Metal drums can be fitted with a petcock and pressure gauge and be safely pressurized up to 2 psi, raising the temperature and shortening the cook time to 8 hours. Marc R. Keith, known on shroomery.org as RogerRabbit, has invented an elegant and inexpensive low-pressure autoclave built out of two metal drums, which is suitable for a small farm or a keen mushroom homesteader. Plans are available on his website (see RR Video in Resources). Using a well-insulated chamber can save much time and energy. For more info on super pasteurization, see "Steam Pasteurization," below.

Pasteurization and Alternatives

Pasteurization

Pasteurization involves heating a substrate to 140–170°F (60–77°C) for 1 hour or more. This kills off filamentous fungi and their spores as well as bacteria, but some bacterial endospores and thermotolerant fungi are spared. The thermotolerant fungi serve as non-competitive placeholders and help to prevent the growth of competitors. Substrates that may be pasteurized include straw, pure sawdust, other agricultural wastes, compost, and faux compost. Because these

---

TIPS FOR SAFE AND EFFECTIVE PRESSURE CANNER USE

- Before use (and before purchase if secondhand), inspect the unit for any cracks, clogged valves, or other faults. Read and follow manufacturer's instructions if available.
- Lubricate metal-on-metal (gasketless) lid connections as needed with a light coating of heavy-bodied oil. Petroleum jelly is traditionally recommended, but I prefer coconut oil. For units with a rubber gasket, inspect the gasket for wear, cracks, and proper placement before each use, and skip the lubricant.
- Put enough water into the PC to last through the entire cook. This is usually between 1–3" (2–6 cm); rocker types lose more water than petcock types. If you are in the middle of a cook and suspect that the water has run out, immediately turn off the heat and let the unit depressurize and cool gradually. A PC run dry can become dangerously hot, breaking jars and potentially rupturing.
- Use a rack or at the very least a cloth to keep the jars or bags off the bottom of the pot. This will help to prevent jar breakage or bags from getting too hot and melting.
- Ensure even heating of the PC and its contents. Attempt to roughly match the temperature of the unit with its contents before loading. For example, if loading hot jars of LC medium, gently preheat the PC with some water in the bottom and the lid loosely on.
- Keep filters dry by covering jar lids with aluminum foil.
- PCs can be loaded with two or more layers of jars or bags. A rack for each layer is helpful to stabilize additional layers and allow steam distribution. Place a heavy ceramic plate on top of bags to prevent them from expanding and blocking the steam valves on the lid.
- Ensure that the lid is properly aligned and evenly seated, then fasten by tightening opposite clamps little by little in a star pattern.
- Once the PC is loaded and lid is fastened, open the petcock valve or keep the weight off of rocker type units. Heat the unit on the stove until steam begins strongly jetting out from the open valve or nipple. Set a timer

substrates tend to have low amounts of free simple sugars and moderate-to-low nitrogen levels, bacterial contamination from surviving endospores is usually not a problem. Several pasteurization methods are discussed below.

The super pasteurizer at Northside Fungi in Enderby, BC. A propane burner heats water in the bottom of the metal barrel, creating steam that is piped into the plastic barrel, where substrates are held on racks. WILLOUGHBY AREVALO

Water Bath Pasteurization for Dry Substrates

This method is good for straw, other agricultural wastes, and random plant materials because it hydrates effectively while washing away spores and bacteria of potential contaminants along with superficial nutrients that may otherwise feed contaminants. However, it

for 5 minutes, allowing the PC and its contents to become thoroughly heated and all air to be expelled. Avoid skin contact with live steam. It can easily and quickly burn you.
- After 5 minutes of bleeding steam, close the petcock or apply the weight in the 15 lb. position. If using a PC with a gauge, monitor the pressure closely as it climbs.
- Once pressure reaches 15 psi or weight begins to rock, lower the temperature enough to maintain even pressure or a steady rocking sound. Start a timer for the duration of your sterilization, which will vary depending on the substrate and vessel size being treated.
- If the pressure climbs above 18 psi, gradually bleed a little steam by pushing the petcock open with a wooden spoon.
- If the pressure drops below 15 psi for more than a moment, restart your clock from the beginning once it gets back up to pressure, to ensure thorough sterilization.
- Do not leave PC unattended during a cook. Getting distracted for even just a few minutes with the stove set too high or low can cause the pressure to become dangerously high or ineffectually low. Whatever you do, don't fall asleep!
- Once pressure has been maintained for the desired time period, turn off the heat and allow the PC to depressurize and cool in its own time. Don't move PC, open the valve, or remove the weight until the unit has depressurized to at most 3 psi. Don't shock the PC by cooling in a cold-water bath or otherwise; if you do, you risk cracking the unit.
- Open the PC when pressure is between 0–1 psi to avoid sucking in dirty air. Do not be impatient. Opening a pressurized PC can lead to steam burns, overboiling of media, broken jars, and wet filters.
- If cooled before opening, the PC will have a vacuum inside. This is fine. To filter out contaminants, drape an alcohol-soaked paper towel over the petcock before opening and equalizing pressure. Open the lid in the lab space if you have one. This isn't necessary if inoculating with LC. Inspect for and discard broken or cracked jars.

▲ Water bath pasteurization in process. CARMEN ROSEN

▲ Slowly lifting out the pasteurized substrate to drain and cool. CARMEN ROSEN

may also decrease yields by removing nutrients. Energy input can be reduced by heating water fully or partially using solar energy or compost heat. You'll need: shredded substrate, permeable bag(s), twine, a big pot or other vessel with a lid, water and a way to heat it, a weight, and a probe thermometer.

1. Load shredded substrate into permeable bags such as burlap sacks, onion bags, or pillowcases and tie shut with twine. For deeper vessels such as a metal drum, a wire basket may be constructed and used instead of the bags.
2. Heat water to 180°F (82°C) in a large pot or 55-gallon drum. Alternately, a bathtub, plastic barrel, or other vessel can be filled with water at this temperature. At the farm where I worked, we filled 150-gallon stock tanks with 180°F water from a modified on-demand water heater. Boiling water can be mixed with hot water from the tap to save heating time. Make sure not to fill all the way; leave room for the substrate.
3. Put the bags or basket of substrate in the hot water, which should cause the temperature to drop to the high end of the desired temperature range. Weight down the substrate with a clean rock, pot of hot water, or other weight to keep it fully submerged.

4. Insert the probe of a meat or compost thermometer into the core of the substrate and cover with a lid to retain heat. Wrap with the vessel with insulative material if possible.
5. Keep temperature between 140–170°F (60–77°C) for at least 1 hour and up to 2 hours. If your vessel is large and well-insulated, you should not have to add more heat. If you must add heat, you could either turn the burner back on for a bit or add a bit of boiling water, distributing it as best you can.
6. Either drain the vessel or lift out the substrate, being careful not to burn yourself with the hot water. Teamwork may be in order. Tying a rope to each bag can aid removal (like a teabag).
7. Allow the substrate to drain until it stops dripping. By this time, it should have cooled enough to inoculate (around body temperature is ideal).

Either discard the water or use it for one more batch. Do not exceed two batches of straw in the water, as excess nutrients leached into the water will build up to a level that is toxic to the mycelium and may promote contamination. This nutrient-rich infusion may be diluted and used as a fertilizer for plants (once cooled) or used full-strength as a biodegradable weed killer.

Steam Pasteurization

Steam pasteurization is ideal for prehydrated substrates, such as unsupplemented (pure) sawdust, compost, or compost-like blends, which have already been packed into bags, trays, or other growing vessels. Like in the hot-water bath pasteurization method, the internal temperature of the substrate is heated to 140–170°F (60–77°C) for 1 to 2 hours, monitored with a remote-probe BBQ thermometer. When the substrate core has reached 160°F (71°C), turn off the steamer, and the temperature will probably continue to climb up to around 170°F (70°C), which is the hottest you should allow the substrate to get. The same equipment and process can be used for super pasteurization; however, the substrate core should be kept at 205–212°F (96–100°C) for 10–14 hours.

> **ROCKET STOVES AND OTHER FOSSIL-FUEL FREE-BOILERS**
> Energy input is one of the biggest negative environmental impacts of mushroom cultivation. The use of biomass-fueled rocket stoves and other efficient wood-burning heaters can offset this. Following his low-pressure autoclave design, Marc R. Keith also designed an amazing high-efficiency, wood-burning boiler; this is also on his website.
>
>
>
> ▲ A fuel-efficient, wood-fired rocket stove at Mushroom Maestros in Lake County, California. WILLOUGHBY AREVALO

Fruiting Substrate Formulation and Preparation • 121

Water bath pasteurization can also be used for prehydrated substrates (like sawdust) or casings—just put them in a jar or watertight PE bag, submerge in a water bath, and weight them down so they don't float. Monitor interior substrate temp just as you would with other methods. Water bath super pasteurization (a 12-hour boil) virtually sterilizes bioactive old woodchips and other coarse plant matter. The addition of a bit of wood ash and <5% hi-cal lime may be beneficial.

By cranking up the temperature on your domestic hot water heater, you can get water into the pasteurization range. DO NOT forget to turn it back down to a safe level as soon as you are done, and use caution as the water left in the tank can remain dangerously hot and cause serious burns! Children are especially at risk if water is too hot!

Many steam pasteurizer designs exist, and it is easy to improvise a good system with a hodgepodge of found materials if you follow the guiding principles:

- The steam chamber should be scaled to your steam supply, growing space, and spawn supply.
- The inside of the chamber should be fitted with racks and spacers as needed to hold the substrate and allow the free flow of steam around each bag.
- The chamber should close tightly and be insulated to minimize the amount of heat loss and energy use.

I have seen people use plastic bins, barrels, coolers, and more. One friend uses a pallet with tarps and comforters wrapped around for insulation.

Steam may be made in the chamber by boiling water in the bottom of a big pot or metal barrel fitted with racks, like a giant vegetable steamer. Stovetop or propane heat is typical, but electric water heater elements are becoming increasingly popular. Alternately, steam can be piped in from an external source, such as an electric steam cleaner, wallpaper steamer, or a pressure canner with a piece of hi-temp tubing fitted in place of the petcock (PC modifications may be dangerous—use caution). The Wagner 705 Powersteamer is a popular model for home growers, as its water tank is big enough to supply steam for 75 minutes. A 9,000-watt sauna steamer is ideal for larger volumes of substrate. These units cost about $200, used. If injecting steam from an external source, be sure to distribute the steam within the chamber with perforated heat-resistant pipe. I have made steamers with both the rack-in-pot method and an external steam source to insulated box, and I prefer the latter for its quicker heating, better heat retention, and seemingly lower energy input.

Solar Pasteurization

In hot, sunny conditions, solar pasteurization is another option that mimics the steam pasteurization method. Using solar energy to heat a sealed and insulated chamber, bags of pre-moistened substrate can reach and maintain pasteurization temperature. This can be done in cold frames, an insulated chamber such as an old fridge with a glass lid, or even in a car parked in the sun on a hot day. As with steam pasteurization, ensure that there is space for airflow around each bag and monitor the internal substrate temperature with a probe thermometer. Upper layers of substrate may dry out when using this method.

Anaerobic Fermentation

Bulk substrates such as straw and woodchips can be fully submerged in water and allowed to ferment, as an alternative to pasteurization (it is sometimes called *cold water pasteurization*). The principle is that underwater, dissolved oxygen will soon be depleted until no *aerobic* (oxygen-loving) organisms, such as fungi, animals, and many bacteria, can survive. *Anaerobic* (oxygen-hating) bacteria will proliferate and consume many of the free sugars on the surface of the substrate along with the drowned aerobic organisms therein. When sufficiently fermented, the water will be very stinky, and an aerobic *biofilm* (scum) will have formed on the surface. This process normally takes between 3 days and a week—or longer in cold weather. The anaerobic bacteria die once thoroughly drained; their stink soon leaves along with their tiny souls, leaving the substrate mostly free of competing organisms. Proceed with inoculation as soon as thoroughly drained.

You can add *Saccharomyces cerevisiae* (ale yeast) as a starter for this process, along with a nutrient source such as molasses or malt extract at a rate of 1–5%. Allow the brew to ferment 2–4 days before submerging the straw for up to 48 hours. This results in a less obnoxious odor.

The resulting wastewater can be applied to cold (slow) compost piles, leaf piles, weeds, or gravel driveways, but not in soil where desired plants are actively growing or into storm drains that lead to creeks. Results are typically not as good with this method as with pasteurization, but the lack of energy inputs and reduced labor and infrastructure can make it worthwhile.

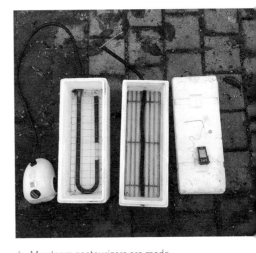

▲ My steam pasteurizers are made from two large Styrofoam fish coolers borrowed from the waste stream. These are easily scavenged behind restaurants or seafood markets. A 1,300-watt steam cleaner I bought at a yard sale pumps steam into one chamber through a perforated washing machine drain tube, and through an old manifold from a gas oven in the other. The racks in the bottom of the coolers and other spacers keep room for steam to pass all around and in-between each bag. Once the first chamber is hot, I move the steam hose to the second chamber, closing up the hole in the first with a cork. I tape the lids down to reduce steam loss. I move the steam hose from chamber to chamber as needed to keep the temperature up. WILLOUGHBY AREVALO

▲ An old fridge loaded with bags of shredded, hydrated ag wastes, and covered with glass for solar pasteurization. This run was a bit overloaded, causing some parts of the substrate to be too hot and others not hot enough. WILLOUGHBY AREVALO

pH Treatment

Straw for oysters can be soaked 12–24 hours in a pH 9.5 hydrated lime (calcium hydroxide) solution (0.5–1% by weight), shifting the straw's pH up to the high end of the oyster's range of tolerance but beyond that of most competitors—as well as most other mushrooms. Use lime that is low in magnesium. Handle and dispose of this caustic liquid with care (wear gloves), and ensure it does not enter soil or a watershed without first diluting or buffering it to a safe pH. Lower concentrations of lime can aid many plants, so when buffered or diluted, the wastewater can be applied to lawns, vegetable gardens, and other plants. Some cultivators love this method.

Hydrogen Peroxide Treatment

Straw and other dry substrates can be soaked for 12 hours in 0.3% hydrogen peroxide (H_2O_2), drained, then inoculated. Fungal mycelia are resistant to its oxidative effects because they produce similar compounds for their own metabolism and self-defense, but it kills bacteria, yeasts, and spores. Biochemist Rush Wayne invented and published methods using H_2O_2 in many parts of the cultivation process to avoid contamination without (see Resources).

Laundry Style

Castile soap or biodegradable detergents containing fatty oils and surfactants can kill most competitors while washing away superficial nutrients that can feed bacterial and mold growth. A washing machine with a mesh bag to hold substrates would save labor, but it may be unpopular with your landlord, roommates, or partner. Add biodegradable detergent, wash, rinse, and spin on hot or sanitize cycle. If it is a top-loader, use 2 or 3 bags to distribute the substrate in the tub so the washer doesn't dance across the room during the spin cycle.

	TREATMENT							CONTAINERS				INOCULUM				
	Sterilization in Pressure Canner at 15 psi	Water Bath Pasteurization	Steam Pasteurization	Anaerobic Fermentation	Hydrogen Peroxide Treatment	Hydrated Lime (pH) Treatment	No treatment except hydration	Glass jars	Polypropylene bags or containers	Polyethylene bags or containers	Other (metal, natural, etc.)	Tissue or spores	LC	Agar	Grain spawn	Sawdust spawn
LC	15–20 min							X	Bottles & jars			X	X	X	X	X
Agar	30 min										Petri dishes	X		X		X
Grain	Prepared and in vessels 1.5–4 hr							X	X			X	X	X	X	X
Sawdust	Prehydrated in vessels, 1–2 hr	Prehydrated in vessels, 1–2 hr	Prehydrated in vessels, 1–2 hr				Can work for very aggressive/naturalized strains	X	X	X	X				X	X
Nutrified Sawdust	Prehydrated in vessels, 2 hr		12–14 hours at 100°C/212°F					X	X	X			X		X	X
Straw/ag wastes		1–2 hours		X		X		X	X	X	X				X	X
Coffee grounds		Prehydrated in vessels, 1–2 hr	Prehydrated in vessels, 1–2 hr				ASAP, within 1–2 days of brewing	X	X	X					X	X
Compost and compost-like blends		Prehydrated in vessels, 1–2 hr	Prehydrated in vessels, 1–2 hr					X	X	X	Trays				X	X
Paper products	If nutrified. Prehydrated in vessels, 1–2 hr	Prehydrated in vessels, 1–2 hr	Prehydrated in vessels, 1–2 hr				X	X	X	X	X	X	Use lots of LC to hydrate	X	X	X
Casing Materials		Prehydrated in vessels, 1–2 hr. Low temp.	Prehydrated in vessels, 1–2 hr. Low temp.				Possible, but increases risk of molds	X	X	X	X					

OUTDOOR CULTIVATION AND MUSHROOM GARDENING

GROWING MUSHROOMS ON LOGS AND STUMPS

The oldest method of mushroom cultivation has stood the test of time and is still the first method many cultivators attempt. However, some people try and fail, or give up hope before their logs have had a chance to fruit. Success with log culture depends on timing, wood selection, spawn quality, and maintaining good conditions for mycelial growth and mushroom development.

Begin by selecting and harvesting good wood. As with sawdust, most wood-loving mushrooms prefer hardwoods without strong antifungal properties. See "The Substrates: Sawdust," in Chapter 7 for a discussion of appropriate woods. If you are curious about the suitability of a wood not mentioned, look into some of the resources in Appendix 2.

Cut logs from live trees in winter to early spring when the sap has begun to flow and before the leaves unfurl; this is when the wood has the highest level of sugars and bark is tightest. Wood cut at other times will work but will yield fewer mushrooms. The ideal size for production logs is 4–6" x 4' (10–15 cm x 120 cm), though other sizes can work. It is said that shiitake logs can produce for as many years as the log has inches of diameter. Cut cleanly and avoid decayed wood, abundant branch stubs, or damaged bark. Store logs out of

▲ A single huge shiitake spontaneously fruits from a neglected alder log, five years after inoculation with the wedge technique. Log-grown mushrooms tend to be especially robust and meaty. WILLOUGHBY AREVALO

direct ground contact. Inoculate logs as soon as possible or up to 8 weeks after harvest before other fungi get a head start on the wood.

Inoculation

Two methods predominate for log inoculation: either plug spawn is tapped into holes with a mallet, or sawdust spawn is inserted into holes with a funnel and dowel or with a palm inoculator (a

▲ Plug spawn is made by soaking and pasteurizing 5/16" spiral-grooved birch furniture dowels and inoculating with sawdust or grain spawn. Plugs are ready to use when thoroughly covered in mycelium. Plug spawn can also be used for spawning beds. Ready-to-use plug spawn can be purchased from suppliers, but beware of old plug spawn that has lost its vigor. OLGA TZOGAS

specialized tool similar to a potato gun, available from mushroom cultivation suppliers). The palm inoculator is stabbed into the sawdust spawn to fill the chamber then plunged into the holes. Plugs are more convenient, but I prefer the higher inoculation rates and cheaper materials of the sawdust spawn method.

Drill holes 1½" (3 cm) deep in a 4" (10 cm) diagonal grid pattern over the wood, approximately 4–6 staggered rows of 8–12 holes. For plugs, use a 5/16" bit, and for palm or funnel inoculation, use a bit with the same diameter as your tool; mine is ½". Larger holes are not recommended because spawn may fall out. If you're going to do a lot of logs, consider buying an auger bit for an angle grinder, which bores holes way faster and easier than a drill. Sanitize hands and tools with alcohol, then get the spawn into the holes, making it flush with the surface of the bark.

> Cutting down trees and operating a chainsaw are inherently dangerous activities. A hands-on training course is highly recommended. Look online for programs in your area.

▲ Plug spawn covered with mycelium, ready to eat this log. MAX BROTMAN

▲ Using a palm inoculator. WILLOUGHBY AREVALO

▲ Applying molten wax to the inoculation sites. WILLOUGHBY AREVALO

Seal the inoculation holes with hot molten beeswax or cheese wax applied with a paintbrush or turkey baster. Sealing keeps out contaminants and fungus-eating invertebrates, and keeps in moisture. Some people wax the cut ends of the logs too, but this is optional and uses a lot of wax. I use a small slow cooker to melt my wax, though a double boiler or a dedicated pot or tin can on a burner work as well. Logs are labeled with species, strain, and inoculation date inscribed on tags that are cut from an aluminum can and tacked on with small nails.

Spawn Run

Stack logs tightly in a shady place out of direct ground contact, like on a pallet. The close quarters promote mycelial growth as well as competitor growth, so after a month or two, restack more openly like a log cabin (crib stack), A-frame, or lean-to to reduce competitor pressure. Monthly watering (or soaking for 2 hours or less) is beneficial in very hot, dry summers. Sufficient myceliation to support fruitings

CHAINSAW STYLES: WEDGES, GROOVES, LAYERS.
Instead of drilling holes, wedges are cut from the logs, cut surfaces packed with a layer of sawdust spawn, and wedges are replaced and nailed or screwed back in place. Or, grooves are cut into logs or stumps and packed with sawdust spawn or myceliated rope. Larger, shorter rounds of wood can be stacked like a layer cake with sawdust spawn as the filling and tacked together with nails. With the wedge, groove, and layer methods, inoculation sites can be wrapped with plastic rather than sealed with wax, and cardboard spawn can be substituted for sawdust spawn.

▲ Claire's log fruits months later, after a good soak. MAX BROTMAN

▲ Shiitake logs incubating in a *crib stack* at Pachamanka, the land stewarded by earthworker and social ecologist Nance Klehm in Illinois. WILLOUGHBY AREVALO.

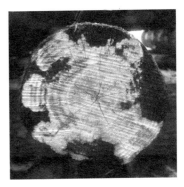
▲ Abundant mycelium on log end shows the pattern of inoculation. This indicates that the log is ready to fruit. WILLOUGHBY AREVALO

◀ Claire Brown packs chainsaw grooves with sawdust spawn. MAX BROTMAN.

typically takes 6 months to 2 years, depending on spawning rate, species and strain, temperatures, wood type, and log size. Mycelium and wood degradation may be visible at the ends. Logs may be a bit lighter than at inoculation. Judging when logs are ready to fruit usually takes experience, but there are a couple of ways to test. A drop in pH from 5.5–6 to 3.8–4 indicates readiness (put 10 g interior wood into 100 ml distilled water and measure pH). One may cut a thin round from the end of a log, moisten it with distilled water and place in a plastic bag. Mycelial growth should be visible within a week if it is ready to fruit. If climactic conditions are conducive and logs are ready, spontaneous fruitings will let you know.

Fruiting

When left to the weather, myceliated logs will usually give 1 or 2 flushes per year. To induce fruiting during warmer months (May–August), logs are soaked in water for 24 hours and restacked upside-down from how they were previously stacked. The inundation of water expels CO_2 that has built up in the wood and provides the moisture for mushroom formation. Shiitake logs are further stimulated by a physical shock, like being slammed on the ground just be careful not to overly damage the bark. Logs should be allowed to rest and dry out for about 7 weeks after a flush is harvested, but can be encouraged to fruit up to three times per warm season.

Growing Mushrooms on Stumps

Stump cultivation is a great laissez-faire approach. Because stumps are still connected to the tree's roots, they wick moisture up from the ground. Stumps are best inoculated when freshly cut, before a wild mushroom flush in the area sends spores galore onto the cut face. They are inoculated in the same way as logs, though the cut face is also inoculated, mostly in the nutritious sapwood close to the bark. Gashes or grooves may also be cut into a stump with a (chain) saw and packed with sawdust spawn. Species particularly suited to stumps include pioppino, nameko, oysters, enoki, maitake, reishi, cauliflower mushroom, chicken of the woods, and turkey tail. Shiitake is not recommended for stumps. If growing gilled mushrooms on logs or stumps, be sure of your identification in case poisonous look-alikes grow spontaneously. Inoculated logs can be sunken into sand in pots or into ground like posts, imitating a stump-like condition.

▲ An "A-frame" of shiitake logs resting in the laying yard after a harvest. WILLOUGHBY AREVALO

MUSHROOM BEDS

Many mushrooms that are naturally found growing from organic matter on the ground do well in a mushroom bed or patch. Good candidates include wine caps, shaggy manes, blewits, nameko, Agaricus, (shaggy) parasols, and elm oysters. Depending on the mushroom's natural substrate/ecological niche, the components of the bed may include fresh woodchips, straw, cardboard, burlap, grass clippings, leaf mold, ag wastes, coffee grounds, re-pasteurized spent mushroom substrates, or leached manure and/or compost. See Appendix 1: Species Profiles for details.

Beds should be inoculated with either pure culture sawdust spawn or naturalized chip spawn, which has already been exposed to the microbial diversity of the world outside the vessel. Naturalized

▲ Mushroom Queen Nina O'Malley in bed with King Stropharias (wine caps). These are at the perfect stage for harvest. Note inoculated logs and floating row cover. CHARLIE ALLER

spawn can be made by inoculating untreated woodchips, sawdust, or other plant materials with sawdust spawn inside a bin, bucket, burlap sack, etc. It can also be transplanted from existing mushroom beds or even wild patches. Grain spawn is not suitable for outdoor use as it would be quickly feasted upon by animals of various stripes. Inoculate at a rate of at least 10–20%, with higher rates leading to quicker and more robust myceliation, and sooner fruitings.

Choose a site that is mostly shaded (though certain mushrooms like shaggy manes and wine caps like some direct sun) and protected from harsh, drying winds. Soil moisture should be high, with plants nearby to provide shade, drop leaves to provide mulch, and

> Spraying or broadcasting lots of thoroughly myceliated LC may work for guerilla spawning in landscaped areas, etc. Spore slurries are another low-tech option. Mix a couple of tablespoons of molasses and a gram or two of salt into a 5-gallon bucket of water. Submerge mature mushroom(s) for 4 hours, agitate, then remove mushrooms. Incubate the slurry at 50–80°F (10–27°C) for 24–48 hours and spray or pour on targeted substrate. Expect inconsistent results.

▲ These zucchinis were transplanted through a wine cap bed of cardboard, woodchips, and straw, into the soil beneath. That summer they both fruited abundantly, and many barbecues were had. MAX BROTMAN

transpire to make a humid microclimate. The fungi will in turn build humus and increase moisture retention, nutrient levels, and overall biological activity in the soil, benefitting the plants. Beds made of compost can be planted in with vegetable starts. Woodchip beds can be laid beneath shrubs as a living mulch, or seeded with a cover crop that will provide nitrogen and humidity to the bed. Squash, corn, Brassicas, berry bushes, and other food plants have been shown to benefit from companioning with certain mushrooms. Further, mushroom beds can be sited in or downhill from livestock areas to filter out and consume nitrogen and pathogenic bacteria from fecal runoff.

If you are building your bed on a slope, soil can be built up to form a berm on the downhill side, increasing the moisture-sinking capabilities of the bed. Beds sunken into the ground on contour with a berm below are referred to as *mycoswales* and stabilize against erosion and contribute to groundwater storage.

Beds are best built during spring and early fall, when temperatures are mild and humidity is high. Freezing temperatures inhibit mycelial growth, so be sure that you make your fall bed early enough that it will have time to myceliate before it gets too cold. In ideal conditions with high inoculation of aggressive mushroom strains, a month of growth can be enough to allow a patch to survive the winter. Once the patch is fully myceliated, it should be triggered to fruit soon, otherwise it will die back. Fruiting can be triggered by watering. If dieback occurs, disturb the mycelium and introduce more fresh substrate, allowing it to continue to grow.

Building a Mushroom Bed

- Begin by digging down a few inches in the area where your patch will be to remove groundcover plants and organic matter, unless this would overly disturb the roots of plants you are trying to support.
- Line the bed with a layer of plain cardboard and/or burlap, with any tape or staples removed.

- Spread a layer of woodchips or other substrate materials on top, about 2–4" (5–10 cm) thick.
- Broadcast a thin but contiguous layer of sawdust spawn or naturalized woodchip spawn over the substrate. It should be well broken up for maximum points of inoculation.
- Add another layer of substrate, and if spawn and materials remain, repeat.
- Finish with a layer of straw, leaves, cardboard, burlap, perforated plastic, floating row cover, or perforated plywood to retain moisture. Final depth of the bed should be between 4–12" (10–30 cm).
- Since many bed-grown mushrooms can benefit from light compaction of the substrates, beds can be made as garden pathways or have some rocks or even mushroom logs laid upon them. Water weekly or as needed in dry weather.
- Expect fruitings by 6–12 months after the bed is made. Fully grown beds can be triggered to fruit by watering. Flushes may persist for up to several years. Patches can be expanded once fully myceliated by incorporating more substrate, extending their lifespan.

Log Raft

A variation of the bed for aggressive, wood-loving species such as nameko, pioppino, reishi, and brick tops (*Hypholoma sublateritium*) is to scar the bark of some freshly cut logs with a chainsaw or hatchet, then lay them down in a cardboard-lined depression, fill in the gaps and cracks with sawdust spawn and woodchips, then cover with mulch. Log rafts tend to have more staying power than woodchip beds, potentially giving robust flushes of mushrooms for numerous years.

▲ A mycoswale installed by my cultivation students as part of an ecosystem-based restoration effort in the Still Creek watershed in Vancouver. VASGEN DEGIRMENTAS

NEXT-LEVEL APPLICATIONS

Substrate (Re)cycling

Myceliated substrates that have finished fruiting serve myriad functions in gardens and living systems. Shallowly buried or just left outside, they may flush again spontaneously. I once got a massive flush of nameko from a heap of crumbled spent blocks. Spent substrates can be used as spawn for beds or logs if there is no noticeable contamination, but their growth will be less vigorous than sawdust spawn that has not fruited, so inoculate heavily. Used as a mulch for garden plants, it contributes to the development of humus. It can be both a food source and bedding material for redworms (compost worms), though they will probably need some additional food to thrive. The resulting castings are full of beneficial bacteria and nutrients for plants. Ruminant animals benefit from the added protein and nutrients in myceliated ag wastes. Spent substrates can also be *pyrolyzed* (burned in the absence of oxygen) to make biochar.

▲ Materials for a log raft: scarred alder logs, fresh alder chips, and elm oyster spawn. WILLOUGHBY AREVALO

▶ Elm oysters fruit prolifically between rows of beans from shallowly buried "spent" straw at Fungi for the People. JA SCHINDLER

HARVEST, PROCESSING, AND USE

WHEN AND HOW TO HARVEST

Mushrooms with a cap and stalk are best picked before the caps expand fully and spore dispersal becomes heavy. Try to pick veiled mushrooms such as nameko, wine caps, and pioppino before or just after the veil breaks. Lion's mane should be picked when the spines begin to elongate. The more mature a mushroom is at harvest, the shorter its shelf life will be. Most mushrooms are best picked with a twisting motion, though some people choose to cut shiitake with scissors close to the base. Pick clustered mushrooms by the bunch rather than individually.

Storage

Refrigerate mushrooms between 35–41°F (2–5°C). Shelf life varies between a couple of days and a couple of weeks, depending on species, strain, water content, maturity at harvest, and fridge temp. Shiitake keep better than most, and oysters don't keep long. Flavor, texture, and nutritional value of mushrooms depreciate quickly in storage, so use them fresh to best enjoy them. Mushrooms should generally not be stored in sealed plastic, as their metabolic processes continue post-harvest, requiring air exchange. Waxed or unwaxed paper bags, plastic baskets, or other ventilated containers are good.

▲ A bouquet of blue oysters ready for harvest. JASON LEANE

BASIC COOKING TECHNIQUES

Sauté

Heat skillet with a healthy (generous) amount of high-quality fat. Add mushrooms when hot but not smoking. Garlic, onion, etc. can be added before, at the same time, or once the mushrooms release some juice, depending how well-done you want those to be. If adding meat or vegetables, add these at the appropriate time so all is done cooking simultaneously. Don't overcrowd the pan, or mushrooms will stew in their own juices and not brown nicely. Good browning increases the umami flavor of mushrooms. Move around by tossing or turning with a spatula now and then, but not constantly. If pan

▲ A mixed mushroom sauté of homegrown nameko, lion's mane, oyster, and wild morels, served with squash and asparagus. WILLOUGHBY AREVALO

dries out, add more fat. Salting early-on will draw out juices. Once mushrooms have browned nicely, the pan can be deglazed with wine, stock, pickle juice, cream, or other liquids. Finishing with a glaze of honey, maple syrup, teriyaki sauce, or other sweet fluid can be delicious. If seasoning with fresh herbs, add these at or near the end. Sautéed mushrooms are delicious with/in just about any kind of savory dish from any cuisine, from quiches to curries, pastas

▲ *Tom Kha* soup with enoki and oyster mushrooms. ISABELLE KIROUAC

to pastries, rice to roasts, tacos to toasts, sandwiches to sushi, and omelets to chawanmushi.

Soups

Many mushrooms are good added directly to soups during the last 5–15 minutes of cooking, depending on the type of mushroom, size of cut, and style of soup. Dried mushrooms can be reconstituted directly in soups; allow longer cooking time. Mushrooms infuse their flavor wonderfully into blended/creamed soups. Perhaps reserve some unblended for texture. Mushrooms can be sautéed before adding to a soup for added umami.

Mushroom Stock

This is a great use for tough stems, woody/leathery medicinal mushrooms, overripe mushrooms, or those that aren't at peak freshness (but not yet spoiled). Gently simmer mushrooms along with any combination of bones, vegetables/veggie scraps (excluding Brassicas and potatoes), herb stems, and spices from 1–6 hours depending on the ingredients. Do not allow to boil hard. Skim away any scum from the surface. Strain and use right away or freeze in ice cube trays for easy portioning and thawing.

Roasting/Grilling

Stuffed or not, whole mushrooms are great when seasoned, oiled, and oven-roasted or barbequed. Mild marinades add flavor and juiciness, but take care not to overpower the mushrooms. Avoid heavy blackening as off-flavors may develop.

Battered and Fried

This is great for large flat caps or cutlets of massive mushrooms. Prepare three plates or shallow bowls with: 1) seasoned flour, 2) seasoned egg (or sourdough starter for vegans), and 3) seasoned breadcrumbs, cornmeal, ground dried myceliated grain, and/or nut/seed meal. Heat skillet with at least ¼" (3 mm) of quality high-heat oil (ghee is supreme). If deep frying, use a thermometer and keep oil around 350°F (175°C). With a fork for each battering stage and tongs for frying, move mushrooms through the stations into the oil, and fry each side until golden. Drain on rack or paper towels. Top up oil between batches if needed, and allow it to heat up before adding more mushrooms. Tempura mushrooms are also excellent.

PRESERVATION METHODS

Dehydration, Rehydration, and Powders

Drying is perhaps the easiest way to preserve an abundance of fungus. Mushrooms can be dried whole or sliced. Use a dehydrator if possible, keeping the temperature moderate if there is a temperature control. Otherwise, mushrooms can be spread out in any way that gives good air circulation and perhaps gentle heat. Racks or screens are ideal, but trays or cookie sheets lined with cloth can also work. Small mushrooms can be threaded into garlands and hung in a warm, airy place. Sun drying is great when weather permits, increasing the nutritional content of the mushrooms by synthesizing vitamin D. Once consistently dry enough to be hard and brittle, store dried mushrooms sealed in glass or plastic. Dry mushrooms can keep for years. Insect infestations occasionally occur, though this is more common with wild mushrooms.

Dried mushrooms can be slowly reconstituted in cool water or other liquid, which preserves more flavor. If in a hurry, hot water can be used. Either way, don't waste the resulting broth, which can be reduced, added back in later or used in another dish. I usually gently squeeze reconstituted mushrooms before sautéing. Dried mushrooms can be added directly into rice (along with a little extra water), soups, stews, and other wet dishes.

Powdered in a spice (coffee) grinder, powerful blender, or mortar and pestle, dried mushrooms become an incredibly versatile seasoning, offering an umami base-note in whatever you want. Meat rubs, sauces, fried potatoes, patties, vegetables, soups, stuffings, popcorn, vinaigrettes, you name it, it's better with mushroom powder. This is a good way to use tough stems and trimmings.

Pickling and Canning

Most firm-fleshed mushrooms pickle well. Mushrooms that are on the edge of spoiling are not suitable and are better dried or frozen.

Many pickling methods and recipes exist, though it is always important that mushrooms are adequately acidulated and/or salted for safe-keeping and the prevention of botulism, which can be deadly.

A delicious Italian method involves boiling mushrooms 5–12 minutes in a brine that is 2:1 good vinegar to water with 1 Tbsp salt per quart of brine and whatever spices and herbs are desired, then draining the mushrooms well (reserving the mushroom-flavored brine for vinaigrette), and packing in sterilized jars submerged in olive oil. Pickled mushrooms can also be hot-packed in their brine and covered with a ½" (1 cm) of oil. These keep refrigerated up to half a year, but should be eaten up fairly quickly once opened. Always ensure the mushrooms are fully submerged in oil, adding more to cover as needed.

Freezing

While a few species of mushrooms, such as shiitake, oysters, and young boletes, can be individually quick frozen raw with pretty good results, most mushrooms are best quickly and simply sautéed before chilling and freezing in portioned freezer bags, with as much air removed as possible. Once thawed, they can be sautéed a few minutes more and incorporated into a variety of dishes.

MUSHROOMS AND MYCELIUM FOR MEDICINE

Following a long history of medicinal use of mushrooms and recent clinical studies supporting this tradition, the home cultivator can produce simple natural medicines. Almost all the edible cultivated mushrooms have been found to contain medicinal compounds, and many inedible mushrooms do as well. Though much variation exists between mushroom species, a few generalizations can be made. Most mushrooms have anti-tumor action by inducing *apoptosis*, and many have immune-modulating properties as well, optimizing the body's immune system to be in a balanced state and encouraging the efficacy of the T and B immune cells without causing unnecessary inflammation. These immune-modulating effects can also act as an anti-inflammatory when that is what the body needs. Many also contain compounds with antiviral, antibacterial, antifungal, and/or antioxidant properties, probably as a function of the fungi's own immune system. Certain mushrooms are also supportive of the liver, kidneys, respiratory system, and other organs. Some have anti-diabetic action, cholesterol-reducing action, and adaptogenic qualities. Many good resources abound for further research into the topic, so see the Bibliography and Appendix 2.

There are a number of preparations to get the medicinal benefits from the mushrooms and mycelium into our bodies, and for the edible species, just cooking and eating them may be as good as any. For those that aren't suitable as food due to texture or flavor, as well as for edible types, mushrooms and/or mycelium can be powdered and encapsulated or incorporated into food, decocted in hot water, tinctured, infused into vinegar, or prepared as homeopathies. Liquid culture broth contains high levels of metabolites and has applications as medicine. Medicinal mushroom elder and herbalist Robert Rogers pioneered the creation and use of *mushroom essences*, which are vibrational medicines for the soul, made similarly to flower essences though under lunar rather than solar influence.

▲ Trifecta of cordyceps encompassed in the heart of reishi; two wild species that have been brought into cultivation at Mush Luv in Charlottesville, Virginia. CHARLIE ALLER

Basic Medicinal Preparations:
Tincture, Decoction, and Double Extract

A tincture is an alcohol or glycerin extract of the more volatile, anti-tumor molecules such as triterpenoids as well as other constituents. A decoction is a hot-water extract, which draws out immune-modulating beta-D-glucan and other constituents. A double extract is a shelf-stable combination of the two and is recommended for most medicinal mushrooms. These extracts can be made with fresh or dried mushrooms, mycelium strained out of liquid culture, or myceliated grain. Dosage can vary depending on the mushroom species, type of extract, and for what it is being taken. Consult an herbalist, natural medicine practitioner, or doctor for guidance. The following recipe is adapted from *The Fungal Pharmacy*, by Robert Rogers.

DOUBLE EXTRACT TINCTURE

Begin with Tincture

1. Measure 1 part mushroom/mycelium by weight to 5 parts strong alcohol by volume (40–95% ABV, stronger is better). Vodka, everclear, and grappa yield a neutral flavor, though whiskey, rum, and other booze can be used. Vegetable glycerin and apple cider vinegar are alternatives, but may not preserve a double extract. If you don't have a scale, use enough alcohol to cover the fungus, as per the folk method.
2. Slice, grind, or otherwise process mushrooms to increase surface area.
3. Submerge fungus in alcohol in a glass jar, cover tightly, and store in a cool dark place for at least two weeks. Shake daily.
4. Using a clean French press, tincture press, potato ricer, nylon mesh sack, or your hands, squeeze out as much liquid as possible. Strain again if desired, and store tightly covered.

Make the Decoction

1. Weigh out the same fungal matter and combine 1:10 weight to volume with water.
2. Simmer uncovered (do not boil) for 2–4 hours or until the volume is reduced to half or less of the original.
3. Strain again, and allow decoction to cool. The solids can now be used for mushroom papermaking.
4. Combine tincture and decoction with a ratio to end up with a finishing alcohol content of 25% or more. If using 95% alcohol, this will be 1 part tincture to 2 parts decoction. With 40% alcohol, use a 3:2 ratio. Label and store in the dark or in brown glass dropper bottles. It should remain good for several years.

IN CONCLUSION
SUBSTRATE FOR THOUGHT—TOWARD FURTHER APPLICATIONS

ONCE A FOUNDATION of mushroom cultivation skills has been established, a world of potential opens up for its applications beyond the production of food and medicine. As the practice of mushroom cultivation becomes increasingly democratized, access increases and the diversity of reasons for and ways of cultivating broadens. As our world and climate rapidly change, novel niches emerge and new problems urgently require new solutions. As the terrestrial Earth's interconnectors, great recyclers, and master chemists, fungi offer great potential for collaboration with humans in realizing these solutions.

MYCOPERMACULTURE

Permaculture Principles and Ethics, and Their Application in Home Mushroom Cultivation

Observe and Interact—This is the basis of the symbiotic relationship when you live with fungi. Pay attention, learn from the fungi, and adapt your practice as needed to fit your living situation. Fungi naturally embody these principles. Follow their lead.

Catch and Store Energy—Make the most of what is available when it is available from your (urban) ecosystem, whether it be substrates, supplies, or helping hands, and preserve the energy for later use.

Obtain a Yield—The harvest is an undeniable motivator. Thank your mycelium for its work, and enjoy and share the fruiting bodies of your labor!

Apply Self-Regulation and Accept Feedback—While caring for your fungi, respect their limitations and your own. Make mistakes and learn from them. Furthermore, act as a steward within your spheres of influence, as your actions have implications on your home, ecosystem, and planet.

Use and Value Renewable Resources and Services—Tap into waste streams to feed your cultivation systems. Scavenge, salvage, and upcycle when sourcing supplies, equipment, and substrates rather than buying new stuff whenever possible. By partnering with producers, everyone serves each other.

Produce No Waste—Choose reusable vessels and supplies, or borrow from existing waste streams (i.e., ice cream buckets). Feed spent or contaminated substrates to the soil or other living organisms. Scale culture, spawn, and substrate growth to each other and to your available time and space.

Design from Patterns to Details—Emulate the exponentially branching pattern of mycelium when designing your cultivation systems and spaces. Extend this to your way of thinking and acting in the world. Integrate the patterns and rhythms of the seasons, climate, and geography. Envision the big picture, then fill in the details.

Integrate Rather than Segregate—Connect with your community and other cultivators to pool resources and effort. Integrate your fungi when growing plants, animals, and bacteria. Allow your children to learn and help with tasks within their abilities.

Use Small and Slow Solutions—Begin modestly and grow within your means. Save spore prints to safeguard genetics for the future. Allow your setup to evolve organically as free or cheap materials and equipment become available.

Use and Value Diversity—Be redundant. Grow multiple species and strains of mushrooms using multiple methods and substrates to make the most of the diversity of substrates, materials, and microclimates available, and to enrich the experience of enjoying the harvests. Use multiple tools, and make each tool serve multiple purposes. Share valuable cultures with other cultivators to ensure their survival.

Use Edges and Value the Marginal—Fungal growth is most vigorous at the edges, and mushrooms often form at the interface of one zone and another. Let the edge be a third space of encounter, collaboration, and exchange. Moderate temperatures in your fruiting space to grow warm and cool strains both at the edge of their range. Tuck equipment and fungi into the edges of your living space.

Creatively Use and Respond to Change—Grow with the seasons. Use what is available when it is available. Ride the wave of mycelial expansion by transferring or moving it as soon as fully grown—but before it slows—to keep vigor at its highest.

Ethics

Earth Care—Avoid making garbage. Minimize non-renewable energy inputs. Collaborate with other natural forces rather than working against them. Acknowledge the agency of all life. Respect the life of soil and water.

People Care—While growing mushrooms in the home, be respectful of your housemates, family, and neighbors. Consider social justice and your privilege in your decisions.

Fair Share—Put ⅓ of your energy toward improving your system, ⅓ toward obtaining yields, and ⅓ to sharing with others.

MYCOREMEDIATION ON A HOME SCALE

Certain fungi can degrade toxic chemicals, capture and move heavy metals, and neutralize pathogenic bacteria. While large-scale projects take great expertise, research, lab tests, and regulatory process, home-scale pollution and waste streams can be addressed with home-scale remedies. Fungi can help treat greywater, pet waste, paper waste, kitchen waste, oil drips from cars, bike rags, urine-soaked diapers, cigarette butts, and more. Vermicomposting certain materials after fungal degradation can further the process. Mushrooms fruiting from these projects should usually *not* be eaten. Mycoremediation must be done responsibly, so as to not inadvertently cause further harm to yourself or the environment. If you want to try working with fungi in this way, please do thorough research and planning before beginning.

▲ Oyster mushrooms degrading and fruiting from motor oil-soaked cardboard. MAYA ELSON

MYCOARTS AND FUNGI AS FUNCTIONAL MATERIALS

Mycelium's unique and dynamic physical properties and patterned growth give it great potential as a natural polymer. Depending on conditions, it can be absorbent or hydrophobic, flammable or fire-retardant, biodegradable or decay-resistant. It is insulative, lightweight, and can have good tensile and compressive strength. Much research has been devoted recently to developing mycelial materials for applications in architecture, furniture, packaging, paper, and more. Artists are often at the leading edge of these explorations—Phil Ross was one of the first artists to sculpt with mycelium and is now innovating fungal leather substitutes with his company Mycoworks.

With basic cultivation skills, sculpting with mycelium just takes thinking outside the box. Molds can be filled with spawn and substrate, myceliated pieces can fuse together, and natural materials can be stitched together by mycelium. There remains much to be explored. My recent and current sculptural work, in collaboration with Carmen Rosen and the fungi, plants, and community of Renfrew Ravine, explores relationships between mycelium and the body, and boosts fungal biomass in the Still Creek watershed as part of a larger restoration and stewardship effort.

Eco-artist, maker, and caretaker Kaitlin Bryson's work with fungi includes a series of remediation pillows that were made as offerings to polluted landscapes. Within the series, each pillow was made with naturally dyed fibers and filled with species-specific fungi to target a site-specific contamination affecting marginalized communities in New Mexico.

The late Miriam C. Rice, author of *Mushrooms for Dyes, Paper, Pigments & Myco-Stix*, innovated working with fungal pigments and fibers, including papermaking with polypores. I have been exploring the potential of papermaking with pureed LC and other cultivated mycelium, which, when incubated before drying, stitches itself back together into a strong, thin material. I have yet to refine any protocols but encourage experimentation. Rarely are excellent pigment-producing fungi cultivated, but some could be. Dr. Seri Robinson, Assistant Professor of Anatomy of Renewable Materials at Oregon State University and art scientist, innovated the cultivation of pigment fungi, including molds, for applications in woodworking, for inkjet printing on textiles, as decking protectants, paint colorants, paper dyes, as well as in batteries and solar cells. Their book and website, "Northern Spalting," are listed under Resources.

COMMUNITY-BASED CULTIVATION EFFORTS

As more people grow mushrooms, more opportunities exist for collaboration, education, and sharing. Community labs, such as Bay Area Applied Mycology in Oakland, California, offer members

▲ *Lingzhi Girl* by Xiaojing Yan. *Ganoderma lucidum* and sawdust substrate. 20" x 19" x 18", 2017. XIAOJING YAN

▲ *Mycelial Connections*, by Willoughby Arevalo, Carmen Rosen, *Pleurotus ostreatus* and community members, 2018. The hands represent two networks of mycelium growing together and fusing. CARMEN ROSEN

▲ *Joining in Acknowledgement (for the farmers, for the soil, for the people)*, by Kaitlin Bryson in collaboration with *Trametes versicolor* and Tewa Women's United. Linen and raw silk dyed with madder root, indigo, sandalwood, rabbit brush, and cochineal; micaceous clay, glass, elm. 2018. Exhibition view. KAITLIN BRYSON

▲ Fruiting Bodies is a living sculpture about the cycles of life, death and regeneration. It is woven and filled with plants from the site, inoculated with oyster mushroom spawn, and sealed with beeswax. It is a collaboration between Willoughby Arevalo, Carmen Rosen, and human and non-human community members. Installed in Renfrew Park, Vancouver, 2019. CARMEN ROSEN

access to facilities, equipment, and exchange of ideas and effort. Community gardens, such as Strathcona Community Garden in Vancouver, BC, offer a great platform for collaborative outdoor mycoculture. Mushroom cultures and spawn are increasingly present at seed and plant exchanges. Community-based mycoremediation projects, such as CoRenewal and Amisacho, team with fungi in the Sucumbios region of Ecuador to address pollution of their lands and waters. How can you engage your community with fungi to increase social and ecological resilience?

APPENDIX 1
SPECIES PROFILES

KEY

Growth Parameters

Fruiting Temperature Ranges
- Cold (40-50°F [4-10°C])
- Cool (50-60°F [10-16°C])
- Warm (60-70°F [16-21°C])
- Very Warm (70-80°F [21-27°C])
- Hot (80°F [above 27°C])

Growth Speed (inoculation to fruiting on processed substrates)
- Very fast (1-3 weeks)
- Fast (3-6 weeks)
- Moderate (6-12 weeks)
- Slow (more than 12 weeks)

Note: While brief descriptions are provided, *please do not rely on these alone for identification of wild mushrooms.* Consult a reputable field guide or online resource; preferably cross-reference several sources. Even better, ask someone who knows. Likewise, purported medicinal properties are based on the work of various researchers. For the sake of brevity, these properties are simplified for this text. Please don't rely solely on these descriptions to medicate yourself or others.

▲ Pioppino at Fungi for the People.
JA SCHINDLER

AGROCYBE AEGERITA (SYNONYMS: CYCLOCYBE AEGERITA, A./C. CYLINDRACEA)

Pioppino (Italian), Black Poplar Mushroom, Swordbelt Fungus, Tea Tree Mushroom (Chinese)

Description: Medium-sized to large. Caps rounded at first then flattened, smooth to wrinkled, dry, rich brown when young fading to cream and paler at the margin. Darker with more natural light. Gills attached to stalk, whitish then turning brown as spores mature. Stalk central, whitish, equal or tapering downward, pliant, clustered, with a membranous ring.

Ecology: Clustered on stumps and snags of poplars and other broadleaf trees in Mediterranean climates. Saprotrophic.

Methods of Cultivation: Indoors in bags or vessels of sawdust. Outdoors on stumps, half-buried logs, log rafts, woodchip beds.

Difficulty Level and Yield: Fairly easy indoors, very easy outdoors. Most strains yield moderately well. Expect two flushes from sawdust, more from stumps and logs.

Growth Parameters: Moderate growth speed, fruiting in cool to warm conditions. Fruits from top surface. May benefit from a casing layer. Can sometimes fully develop inside closed bag.

Medicinal and Nutritional Properties: Anti-inflammatory, antibacterial, antiviral, cancer-inhibiting, spleen support. 27% protein dry weight.

Culinary Value and Use: Firm texture holds up well to cooking. Flavor is nutty, meaty, and umami due to high levels of naturally occurring MSG-like glutamates.

Comments: Be sure of your identification if growing on logs or stumps, as poisonous look-alikes may volunteer.

COPRINUS COMATUS
Shaggy Mane, Shaggy Ink Cap, Lawyer's Wig, Maotou-Guisan (Chinese)

Description: Easily recognized medium-sized to large mushrooms with a uniquely shaggy and cylindrical cap, white with brown shags. Gills densely packed (like pages of a book), free from stalk, white when young and turning vinaceous-pink to ink black as mushroom autodigests from the bottom up to aid spore dispersal. Stalk white, hollow, tapering upward with a bulb at the bottom and a ring above the bulb.

Ecology: Decomposer of nitrogen-rich organic matter. Widespread and common on rich soils (lawns, pastures) or in compacted/disturbed ground (paths, gravel roads), even found breaking through asphalt. Fruiting in scattered groups, often prolifically.

Methods of Cultivation: Indoors in trays or shallow bags of pasteurized substrates such as compost, straw, ag wastes, paper, spent oyster substrate, manure, or blends, covered with a casing layer. Outdoors in beds of substrates just mentioned or sawdust inoculated with sawdust spawn. Cover with perforated tarp or board while myceliating, then cover with a casing soil (like potting mix) to encourage fruiting. Mix of sun and shade preferred. Expect flushes in fall or sometimes spring.

Difficulty Level: Difficult indoors (expect low yields), but fairly easy outdoors (variable yields)

Growth Parameters: Fruits 6–12 months after inoculation outdoors, fruiting in cool to warm conditions. Fruits only from horizontal (top) surfaces.

Medicinal and Nutritional Properties: Antibacterial, antifungal, antitumor (especially breast and prostate cancers), blood sugar moderator, 25+% protein, 59% carbs dry weight.

▲ A local strain of shaggy mane fruiting from a cardboard box filled with homemade hot compost, cased with homemade biochar, and planted with grass at Fungi for the People.
JA SCHINDLER

Culinary Value and Use: Succulent and tender, loaded with savory, classic mushroom flavor. Great sautéed with garlic and butter, served with eggs, added to soups, or stuffed, breaded, and fried.

Comments: This mushroom must be picked young and enjoyed promptly, keeping two days max in the fridge before it turns to ink. The ink, if evaporated to concentrate, is lovely for drawing and calligraphy. Shaggies hyperaccumulate heavy metals such as arsenic, mercury, and cadmium and have a fondness for polluted sites, so exercise restraint and prudence when harvesting in the wild. Mycelium predates on nematodes, presumably for their nitrogen.

CORDYCEPS MILITARIS
Caterpillar Fungus, Keeda Jadi (India)

Description: Small orange Ascomycete mushroom. Club-shaped with pimply bumps on the upper portion, which are the spore-producing bodies.

Ecology: Fruiting in summer and fall. Parasite of pupae and larvae of various moths and butterflies. Widely distributed but infrequent in western North America, more common further east. I've only found it twice.

Methods of Cultivation: Indoors on various sterilized, protein-rich substrates inoculated with liquid culture, often grain-based blends supplemented with potato broth, sugar, starch, yeast, and vitamin B1. Substrates based on insects such as superworms, silkworms, or hornworms also support fruitings, as can boiled eggs. Fruiting is best done in shallow trays inside sealed PP bags or in closed wide-mouth jars filled with 1" (2–3 cm) of substrate.

Difficulty Level: Slightly challenging. Indoors only.

Growth Parameters: Fast-growing with spawn run complete in 1–2 weeks, but allow 4 weeks or more for fruit to develop. Keep warm for

▲ Homegrown cordyceps at Mush Luv in Charlottesville, Virginia. The strain was cloned from a local wild specimen. NINA O'MALLEY

incubation and fruiting. Incubate in complete darkness and trigger to fruit by introducing light.

Medicinal and Nutritional Properties: A medicinal powerhouse that rivals the extremely expensive and increasingly rare, wild-harvested *Cordyceps sinensis*. One of the most wide-spectrum medicinal mushrooms with antioxidant, antimicrobial, antitumor, immune-modulating and moderating properties, among others. It tones the kidneys, liver, and nervous, respiratory, and reproductive systems, reduces blood-cholesterol and blood sugar, and restores depleted energy by boosting ATP production.

Culinary Value and Use: Perhaps surprisingly, this mushroom is delicious, bursting with a strong, sweet aroma and mushroomy flavor, and its color and form lend itself well to beautiful presentations on the plate. I think it has great potential to be a gourmet trend food once production takes off in North America. I've fried it in butter until crispy and served it as an accent atop other foods. The substrate block is also edible and medicinal, with great potential as a functional food product. It is often tinctured or encapsulated.

Comments: Small-scale cultivation is just beginning to proliferate in North America. The techniques for growing it have mostly been developed in China, Korea, Thailand, and India. William Padilla-Brown of MycoSymbiotics has published an ebook on small-scale Cordyceps farming, see Resources.

FLAMMULINA VELUTIPES AND ALLIES
Enoki, Enokitake (Japanese), Velvet Shank, Golden Needle Mushroom (Thailand), Winter Mushroom, Snow Cap Mushroom

Description: Wild or natural-grown form is small to medium-sized, with rounded, sticky, peelable, yellow to golden-brown caps, attached gills, no veils, and a white spore print. The flesh is thin. Stalk is

▲ A local strain of enoki called "hairy long legs" cultivated at Mush Luv. These grew in somewhat elevated CO_2 and normal light conditions. Bag was removed for photo. NINA O'MALLEY

▲ The wild cluster of "hairy long legs" enoki, from which the strain was adopted by Charlie and Nina. CHARLIE ALLER

▲ A wild fruiting of "Yukon golden needle" (*F. populicola*) from roots in frozen ground in downtown Whitehorse, Yukon Territory, in late November. WILLOUGHBY AREVALO

▲ Ja, Isabelle, and I found these wild enokis on Doug fir on the trail to McCredie Hot Springs in Oregon late one night. JA SCHINDLER

fibrous to tough, velvety, dark brown to black below and pale yellow at the top. Indoor cultivated forms can differ radically, with clusters of long, thin, smooth stalks, tiny caps, and pale coloration.

Ecology: Mostly saprotrophic, occasionally weakly parasitic. On stumps, roots, logs, and dead branches of various broadleaf trees and shrubs, especially willow, elder, aspen, maple, bush lupine, etc., occasionally on Douglas fir. Often arising from gaps between wood and bark.

Methods of Cultivation: Indoors on nutrified sawdust in jars or narrow bags, also on paper products. Encourage top-fruiting. Outdoors on stumps, partially buried wood rounds, or log rafts.

Difficulty Level: Fairly easy indoors with moderate yield, easy outdoors with low yield.

Growth Parameters: Fruiting temperature is strain dependent, but cold strains can withstand temperatures fluctuating above and below freezing. I've also seen it growing wild in a very warm Quebec summer. Fast-growing indoors. The exaggerated, noodley, pale indoor form is achieved by applying a collar around jar mouths or leaving high walls of bag to pool CO_2, and fruiting in dark or low-light. Harvest noodle-form by the clump and expect another flush or two, a couple of weeks apart.

Medicinal and Nutritional Properties: Antitumor, immune-modulating, anti-allergenic, antimicrobial, anti-inflammatory, and antiviral. Up to 31% protein dry weight, depending largely on substrate.

Culinary Value and Use: One of the sweetest-tasting mushrooms; this is one of the few I like to eat raw (especially the natural form). It should be cooked either very lightly (tossed in soups just before serving) or very well (fried to a crisp), otherwise the stalks are very stringy. Mincing solves this problem.

Comments: Enoki is great for science experiments because of its shapeshifting power. *Be sure of your identification if growing on logs or stumps, as poisonous look-alikes may volunteer.*

GANODERMA LUCIDUM (SENSU LATO), *G. TSUGAE* AND ALLIES

Reishi (Japanese), Lingchi/Lingzhi (Chinese for Mushroom of Immortality), Varnished Conks

▲ A local reishi strain (*G. sessile*) fruiting from a windfall log just two months after inoculation, at Mush Luv in Virginia. CHARLIE ALLER.

Description: Medium-sized to huge annual polypores. Kidney-shaped cap is shiny (varnished), zoned and colored red, orange, golden, or even purplish, brown, or black, paler near the margin, often dusted with brown spores. Pores on underside of cap are tiny, round, and whitish, staining pale brown when scratched. Flesh is brown and corky. Stalk attached to side of cap, sometimes nearly absent but at other times very prominent, depending on environmental conditions.

Ecology: On snags, stumps, and logs; on butts of unhealthy living trees of a wide variety of woods (certain reishi species have affinities for certain woods; others are generalists). The species complex exists nearly worldwide in temperate to tropical climates. I've greeted reishis on old-growth hemlock snags in Cascadian rainforest, on a palm at the beach in Senegal, on a giant oak with a wild beehive in its decay column in Mexico, and also growing from roots of unhealthy sidewalk trees in San Francisco, Houston, Traverse City, and DC.

Methods of Cultivation: While generally grown indoors in bags or jars of nutrified sawdust, it can adapt to a wide range of substrates including coffee and shredded ag wastes. Outdoors it does well on logs, especially when partially buried. Short logs are often potted in sand and fruited in greenhouses.

Difficulty Level: Fairly easy indoors, easy outdoors. Good yields, potential for second flush a couple of months later.

Growth Parameters: Fast to myceliate, but primordia formation and fruiting body development takes weeks to months. Stalk development and capless antler forms can be achieved by boosting CO_2 levels/restricting fresh air exchange, though yields will be lower than with capped/shelving forms. Most strains like it warm, some very warm. Can top or side fruit.

Medicinal and Nutritional Properties: One of the most loved and best studied medicinal fungi. It is one of the most broad-spectrum natural medicines, boasting adaptogenic, immune-modulating, anti-inflammatory, antitumor, anti-allergenic, antiviral (hepatitis, HIV, herpes, flu), cholesterol-reducing, hypertension-reducing, and antimicrobial properties. It tones the liver, kidneys, and nervous, respiratory, cardiovascular and reproductive systems. Contains significant levels of vitamin D, especially when grown or dried with natural light.

Culinary Value and Use: While too tough to chew and digest, my Chinese co-worker and friend, Sally, cooks chicken with dried shrimp in Lingzhi broth. Reishi-infused vodka is a go-to element in many of my improvised cocktails, with its earthy bitterness complementing spicy, sweet, fruity, and bright elements. I also occasionally homebrew a reishi-ginger-turmeric pale ale, originated by my friends Max Brotman and Claire Brown. Find our recipe in the book *Radical Mycology*.

Comments: Due to its beauty, tenacious mycelium, and aggressive growth, reishi collaborates well with humans on sculpture and functional materials projects. The taxonomy of the *G. lucidum* complex is a confusing mess, with many similar species and strain-to-strain variation within each. This has implications for cultivation as well as medicinal and form/structural properties. I recommend cloning a wild strain from your bioregion, and/or getting a rad strain from other cultivators. Reishi loves living in LC but is hard to clone with the needle. Try to get tissue from the tender, young margin, or do an agar-to-LC transfer.

HERICIUM ERINACEUS, H. CORALLOIDES, H. AMERICANUM, H. ABIETIS AND ALLIES

Lion's Mane, Pom du Blanc (French), Yamabushi-take (Japanese for Mountain Priest Mushroom), Monkey Head, Bear's Head, Coral Tooth

▲ A backyard fruiting of lion's mane from nutrified sawdust inoculated with liquid culture. WILLOUGHBY AREVALO.

Description: A milky waterfall, frozen in midair. The most graceful muppet. Clusters or branches of white, creamy, or pinkish, icicle-like spines that lengthen with age. Flesh is the same color, often filled with air spaces, slightly elastic, fragrant. Medium-sized to huge. Spores white. Easy to ID to genus, all members of which are good to eat.

Ecology: Facultative (weak) parasites, fruiting in summer and fall from wounds or woodpecker holes in living trees, or saprotrophic on recently dead wood. *H. erinaceus* prefers oaks, *H. americanum* maples, and *H. abietis* firs. Locally abundant in certain ecosystems but generally rare. *H. erinaceus* is red-listed in 23 countries, making propagation of local strains extra relevant.

▲ Olga Tzogas and her local *Hericium americanum* strain in the mist. WILLOUGHBY AREVALO.

Methods of Cultivation: Generally grown indoors on sawdust, but it has been grown on various ag wastes, too. Often grown in bags with one or more ¼" holes poked through side of bag at initiation for fruiting sites. Hericiums will also fruit readily from grain or culture media. Outdoors, it is grown on logs, Shiitake style, though papery-barked woods like alder and birch may not yield fruit.

Difficulty Level: Fairly easy, moderate to good yields.

Growth Parameters: Very fast to grow and fruit. Some strains will form primordia even before myceliation is complete. Most strains like it cool—cold shock if primordia aren't forming on their own. Can flush multiple times. Humidity requirements for fruiting is lower than for pinning. Pick while still firm, when spines are about 1 cm long.

▲ The lengthening spines on this lion's mane show that it's ready to be picked. JASON LEANE

Medicinal and Nutritional Properties: Remarkable for their stimulation of nerve growth factor. Research suggests that Hericiums can help with various neurological problems such as Alzheimer's, Parkinson's, dementia, etc., as well as anxiety and symptoms of menopause; they can also improve general cognitive function. Also immune-modulating, antitumor, and they are used in traditional Chinese medicine for gastric ulcers. Up to 30% protein dry weight.

Culinary Value and Use: Unique yet versatile, Hericiums are rightfully compared to crab or lobster in both texture and flavor. Often juicy, they are excellent sautéed in butter with garlic, basted with olive oil and grilled, or added to soups (like tom kha) 10 minutes before serving. Some chefs slice them, but I almost always tear or leave whole (if small) to preserve their character.

Comments: Nina and Charliceps of Mush Luv advocate the smoking of this mushroom. I found it has a pleasant flavor and gives a clear, calm euphoria.

HYPSIZYGUS TESSULATUS (H. MARMOREUS)
Shimeji, Beech Mushroom

H. ULMARIUS Elm Oyster/Garden Oyster

Description: Robust, medium-sized to large mushrooms with rounded, white or grey to brown caps, shimejis have tessulated "water spots," like cracked glass. Margin incurved when young. Gills broadly attached to central stalk in shimejis, but running down an off-center to lateral stalk in elm oysters (see Comments). Stalk often curved and whitish, with no veils. White spores. Growth is clustered.

Ecology: Weakly parasitic or saprotrophic on elms, beeches, cottonwoods, and Manitoba maples (among others), producing a cubical brown rot in the wood, which is highly beneficial to soil. Often fruiting high up in trees from branch stubs or split bark.

Methods of Cultivation: Typically grown on sawdust or finely shredded ag wastes, usually in bottles or jars. Adaptable to a variety of (dense) substrates. Grown outdoors in sawdust and woodchip beds, or on partially buried logs. Garden oyster grows aggressively and is so-called because it partners well with many garden vegetables, especially Brassicas, and grows well as a living mulch layer.

Difficulty Level: Easy, often high-yielding.

Growth Parameters: Fast-growing, cool to warm fruiting. Fruits well from top or side surfaces.

▲ Beautiful brown shimeji ready for harvest at All the Mushrooms.
JASON LEANE.

Medicinal and Nutritional Properties: Antitumor, antibacterial, antifungal. Low in protein compared to others.

Culinary Value and Use: Shimejis are highly prized for their firm, crunchy texture, rich flavor and presentable appearance. Elm oysters are often regarded as superior in flavor and texture to other oysters. Both are highly versatile and good pickled or dried.

Comments: This group is plagued by taxonomic confusion, especially the concept of *H. ulmarius* among cultivators. Somehow, the cultivated "elm/garden oysters" have lateral stalks and decurrent gills, just like a *Pleurotus*. In contrast, wild elm oysters have gills that are attached but not decurrent, scales on cap and stalk, a mealy rather than sweet smell, and they grow in groups of two or three rather than big clusters. I suspect the cultivated and wild varieties are different species. Because virtually all elm oyster cultures available are this way, and virtually all cultivation books define them in this sense, I reluctantly perpetuate this classification, referring to them here in the context of cultivation. However the wild form can likely be cultivated by similar methods and is worth trying.

LENTINULA EDODES
Shiitake (Japanese), Shiangu-gu (Chinese for Fragrant Mushroom)

Description: Medium-sized to large mushrooms with brown caps, at first tightly rounded with inrolled margins, then expanding to flat, sometimes with a broad *umbo*, or hump in the center. Caps may be decorated with zones of white fur, especially near edge, depending on strain and temperature. Caps may be cracked in dry conditions (especially certain *donko* strains)—Gills buff or reddish-tinged, close, broadly attached to stalk, and with finely serrated edges. Stalk pale reddish-brown, short, tough at base, often curved, equal thickness or tapering downward, and coated with fuzzy fibrils. Flesh white or reddish-tinged, fibrous, aromatic. Spore print white.

▲ Elm oysters (cultivated form) on a column of pasteurized straw/sawdust mix in an outdoor fruiting chamber at Fungi for the People. JA SCHINDLER.

▲ Wild (true?) elm oysters on *Acer negundo* at Pachamanka in Illinois. WILLOUGHBY AREVALO

▲ Shiitake fruiting from nutrified sawdust. Humidity tent was removed for photo. JASON LEANE.

▲ Shiitake mycelium blistering and starting to brown, eight weeks after inoculation. WILLOUGHBY AREVALO.

▲ Shiitake mycelium about a week later, almost fully browned, about to pin. WILLOUGHBY AREVALO

Ecology: Saprotrophic on the shii tree (chestnut oak) and other hardwoods. Native to East Asia and only recently and rarely found naturalized in North America.

Methods of Cultivation: Contemporary sawdust-bag and antique log techniques were created largely with and for shiitake, though other waste streams have produced yields. Between myceliation and pinning, mycelium (for most strains) forms bumps (called *blistering*) then undergoes a 2-week *browning* phase, where increased gas exchange and the red-brown metabolites induce the formation of a brown, bark-like coating over the sawdust block. I like to pull the bag away from the block without opening it, allowing the flow of metabolites and oxygen. One browned, the bag is removed completely, allowing fruiting all over top and sides (even bottom) of the block. For this reason, jars are not recommended, though reusable buckets may work—just slide out the blocks once browned.

Difficulty Level: Easy outdoors, slightly challenging indoors. Typically high-yielding.

Growth Parameters: Many strains exist, generally categorized into warm, cool, or wide-range fruiters, allowing year-round production. Some strains are better suited to either logs or to sawdust. Relatively slow growing, logs may take over a year before fruiting, and sawdust blocks between 7 and 12+ weeks. Harvest flush with substrate with scissors or sharp knife.

Medicinal and Nutritional Properties: Much research has established shiitake as immune-modulating, antiviral (herpes, HIV, flu, etc.), inhibitory of many cancers, anti-inflammatory, antibacterial, antifungal, and moderating of sugar, cholesterol, and pressure levels in the blood. It tones the kidneys and reproductive system. Roughly 20% protein dry weight, containing almost all the essential amino acids. Very high in vitamin D, especially when grown or dried in natural light.

Culinary Value and Use: One of the favorite edible mushrooms of humans worldwide; its unique, potent flavor and aroma are nearly indescribable. Infrequent *mycophagist* (mushroom eater) Paul Kroeger said it best: "It's as if I'm licking the armpit of the Buddha." Its sublime musk and meaty texture find a home in many cuisines and preparations, not only the Asian ones. Excellent fresh or dried.

Comments: If planning to do lots of logs, see Cornell University's "Best Management Practices for Log-Based Shiitake Cultivation Handbook" (listed in Resources). Due to its potent white-rot enzymes, spent blocks may be applied to polluted sites to detoxify PCPs, PCBs, and PAHs.

PHOLIOTA NAMEKO AND ALLIES Nameko

Description: Medium-sized, with rounded, scaly, orange-brown caps covered in a thick slime that forms a translucent orange veil over the young pale gills, which darken as the dull-cinnamon brown spores develop. Cap becomes flat or wavy at maturity and less bright orange. Stalk central, fleshy, and pale with orangish scales and a glutinous ring. Growth is somewhat clustered.

Ecology: Saprotrophic on stumps of oaks, beeches, and other hardwoods in temperate Asian forests.

Methods of Cultivation: Indoors in bags or jars of nutrified sawdust, or outdoors on partially buried hardwood logs, log rafts, or stumps. Nutrified conifer sawdust will also support fruitings. Spent sawdust from indoor growing can fruit prolifically when piled up outdoors.

Difficulty Level: Fairly easy indoors; easy outside, offering abundant harvests.

Growth Parameters: Nameko likes cool and very wet fruiting conditions. I've seen blocks fruiting happily in bags half-filled with standing water. Growth is fast. Primordia emerge from a slimy orange "marmalade" that develops on exposed surfaces about a week after initiation

▲ Isabelle Kirouac with nameko at Mycality Mushrooms in Arcata, CA, about two days before harvest. WILLOUGHBY AREVALO.

▲ Nameko at the perfect stage for harvest. WILLOUGHBY AREVALO

▲ A local strain of Chestnut Mushroom fruiting from nutrified sawdust at All the Mushrooms, a micro farm in Powell River, BC.
JASON LEANE

into cool, airy, and very humid conditions. Bags should be cut open, leaving a few inches of wall standing up around the top of the block, gathering moisture. Mainly a top fruiter, but may side fruit too. Jars can be fruited upright or on their sides. Pick just before (or just after) veils break for best texture and storage life. Trim bases before packing, being careful to keep sawdust out of the slime. Roughen up surface of substrate to encourage pinning for second flush.

Medicinal and Nutritional Properties: Antitumor, anti-inflammatory, antiviral, antibacterial (especially against coliforms). About 20% protein dry weight.

Culinary Value and Use: Don't let the slime psych you out. This is a seriously delicious mushroom with a cashew-like flavor and firm, crunchy texture. When pickled or added directly to soups (toward the end of cooking), they retain their character, and the slime gently thickens the broth (as per the Japanese tradition). When sautéed or roasted until caramelized, very rich umami flavors develop, and the slime disappears. Nameko me hungry!

Comments: I've pondered the potential of the "marmalade" slime as a natural sex lube. Please report back if you try it.

The Chestnut Mushroom (*Pholiota adiposa*) is a similar but scaly and less slimy species native to North America. Its flavor isn't equal to nameko, but this gorgeous mushroom is worth playing with, especially if you can catch a local strain. Same goes for the slender *Pholiota (Kuhneromyces) mutabilis*. Be sure of your ID if growing outdoors, as many brown-spored wood-lovers are poisonous, and many similar-looking *Pholiota* species are unpleasant tasting.

PLEUROTUS OSTREATUS Oyster Mushroom, Pearl Oyster

PLEUROTUS PULMONARIUS Phoenix Oyster, Chocolate Oyster

PLEUROTUS POPULINUS Poplar Oyster

PLEUROTUS DJAMOR GROUP Pink Oyster, Flamingo Mushroom

PLEUROTUS CITRINOPILEATUS Golden Oyster, Yellow Oyster

PLEUROTUS ERYNGII King Oyster

Description: Medium-sized to large, usually clustered mushrooms. Caps at first rounded then uplifted and wavy. Cap color variable depending on species, strain, and environmental conditions, ranging from white to cream, brown, grey, blue, yellow, or pink. Gills running down stalk, colored like caps or paler, often whitish. Stalks often short in relation to cap (though longer in high CO_2 conditions), attached off-center or to side of cap, usually thinner at the base, often with hairy mycelium at base, sometimes tough. No veils. Spore print whitish, lilac tinged, or colored like caps. Odor of mushrooms and mycelium often sweet, like anise. King oysters have a stouter stature and big firm but tender stalks.

Ecology: Saprotrophic on a wide variety of mostly broadleaf woods, fruiting high up in snags or on fallen logs, or sometimes on stumps. Mostly on alder here in the PNW. Phoenix oysters often found (and cultivated) on conifers. Yellows, pinks, and some white strains are tropical in origin. *Pleurotus* species are common and widely distributed. King oysters are unique, being root parasites of the thistle-like Mediterranean Eryngo plant, but they can be cultivated similarly to other oysters.

▲ My Moroccan oyster strain (*P. pulmonarius*) fruiting from an oak stump in Eugene, OR. JA SCHINDLER

Methods of Cultivation: Versatile and adaptable to a wide variety of methods, substrates, and vessels. Most common are straw in perforated bags or buckets, sawdust in bags or bottles, and logs. Well suited to growing on coffee grounds, ag wastes, and paper products. Trays, woodchip beds, laundry baskets, stumps, and other methods are used as well. Side (preferred) or top fruiter. King oysters yield better off of sawdust than straw, and prefer to top-fruit, jars working

▲ Phoenix oysters fruit from coffee grounds in a native plant greenhouse. DANIELLE STEVENSON

King oysters on nutrified sawdust.
JASON LEANE

Pink oysters on nutrified sawdust.
JASON LEANE

well. A casing can improve pinning and yields for kings. In many farms, primordia are pruned to select the best one or two to get big.

Difficulty Level: The easiest mushrooms to grow, indoors or out. High yield potential.

Growth Parameters: Very fast-growing, usually 2–4 weeks from inoculation to fruiting on most substrates. Different species/strains are suited to different temp ranges from cool to hot, allowing for year-round oyster fruiting with minimal temperature control. Stronger light and cooler temperatures increase cap pigmentation. Local strains are easy to catch by tissue culture or stem-butt to cardboard culture. Commercial strains easily obtained from grocery stores and cloned.

Medicinal and Nutritional Properties: Reduces blood-cholesterol and hypertension. Also antitumor, antiviral, antibacterial, antioxidant. Tones the nervous and respiratory systems. High in protein, up to 30% dry weight, also high in potassium and vitamin C.

Culinary Value and Use: Oysters are versatile, adapting as well in the kitchen as in the lab. Younger oysters can be lightly cooked, while older ones are better sautéed until browned or breaded and fried. Decent dried and reconstituted, perhaps better sautéed and frozen, also good pickled. Kings are particularly meaty and delectable, yellows sweet, and pinks seafood-like.

Comments: Oysters are the best choice for novice cultivators, educators working with kids, or for vegetable farmers wanting to add a mushroom crop to their fresh sheet. Their potent enzymes are capable of breaking down many toxic chemicals, so they're engaged often in mycoremediation. Oysters are prolific spore producers, and the spores irritate the lungs, so harvest before full maturation, especially if fruiting in your living space.

STROPHARIA RUGOSO-ANNULATA
Wine Cap, King Stropharia, Garden Giant

Description: Large to huge mushrooms, up to 5 lbs. (2 kg) each. Caps rounded then flattened, eventually upturned; red-brown with whitish furry tufts near edge when young, then fading, with a metallic sheen and cracked in dry weather. Gills start pale, becoming dark purple-black like the spores when mature. Young gills are covered by a thick membranous veil that is split into a cogwheel pattern, leaving a ridged double ring as the cap expands. Stalk is whitish, stout, and thicker at the base, where stringy, white *rhizomorphs* (mycelial cords) are attached.

Ecology: A ripping decomposer (saprotroph) of woody or cellulose-rich debris in the soil. Sometimes found in hardwood forests, but more commonly in human-landscaped areas laid with woodchips, compost, beauty bark, or other mulch. Also occurring in gardens or arable fields; I once found it in a horse pasture. It thrives when living in community with bacteria, and is often weak in pure culture. Disturbance stimulates aggressive growth.

Methods of Cultivation: Wine caps are usually grown outdoors in partial shade, in beds of woodchips, straw, or dried out, chopped stalks of garden plants like corn, sunflower, raspberry, hops, etc., moistened and inoculated with sawdust spawn or myceliated substrate from another bed in the spring or fall. An optional plastic sheet on top maintains humidity and boosts CO_2, aiding myceliation for the first month or two. Mushrooms often grow at the soil interface, so once myceliated, remove plastic and cover with a layer of moist soil to promote fruiting (optional). Squashes, corn, and other garden plants can be transplanted through bed into the soil below, to the benefit of both plant and fungus. Wine caps can also be grown in wooden boxes, large containers, or bags of spent substrate from other mushrooms, straw, or other pasteurized substrates. Sawdust

▲ A basket of backyard-grown wine caps. OLGA TZOGAS

spawn is the ideal inoculum. Once myceliated, cover with a couple of inches of moist casing soil. Peat pasteurized for 60 minutes at max 140°F (60°C) works, as can potting soil straight out of the bag or even garden soil. Mist heavily to encourage pinning, then back off once they start to enlarge. Harvest when veils have just broken or before, as mature mushrooms are of lower quality.

Difficulty Level: The easiest mushroom for growing in beds and gardens, where it can be massively productive. Challenging to grow indoors, and yields are lower too.

Growth Parameters: Mycelium cottony and slow in pure culture; shake grain jars to stimulate. Fruiting in warm weather, usually late-summer to early fall; beds laid in fall fruit in spring.

Medicinal and Nutritional Properties: Under-analyzed. Likely antimicrobial. 22% protein dry weight.

Culinary Value and Use: A summer mushroom with meaty texture and potato-like flavor; good served with vegetables. It is great barbecued, served portabella-burger-style, or sliced thinly and fried. Despite its prolific fruitings, it should be eaten in moderation (not more than two days in a row), as it may cause indigestion and nausea when heavily feasted upon. Fortunately, caps dry well (stalks can be powdered), and it can be frozen. Cook thoroughly.

Comments: At my old garden, I used my Stropharia patch like a garbage dump for coarse garden waste, tree prunings, pizza boxes, etc., and I broadcast my greywater over the patch, hoping the mycelium would eat any nasty bacteria and cooking oils. The patch thanked me for the gifts with abundant fruitings, though slugs and woodlice often got to the mushrooms before I did. They are such shredding decomposers that they will quickly exhaust their substrate, so patches require refreshing every year or two. Worms, centipedes, and other soil invertebrates ride in on their coattails, soon rendering the bed ready for planting vegetables. For gardeners needing to boost organic matter in their soil quickly, this mushroom is a superhero. They have been employed as a biofilter for waterways polluted with coliform bacteria.

TRAMETES VERSICOLOR
Turkey Tail, Coriolus, Yun Zhi (Chinese), Kawaratake (Japanese)

Description: Overlapping shelves of leathery semicircular, medium-sized brackets. Upper surface banded with concentric zones of various colors, including creamy white, browns, copper orange, bluish grey, reddish, and green. Underside white to cream or pale grey, covered with tiny round pores. Flesh tough and fibrous, white. Odor sweet and mushroomy. Spore print white.

Ecology: Worldwide from temperate to tropical zones, extremely common and abundant on logs, stumps, snags, and branches of a wide variety of woods, mostly broadleaf but sometimes on conifers (I once even found it on cedar). Saprotrophic, but often inhabiting dead branches in live trees.

Methods of Cultivation: Indoors on (plain) pasteurized sawdust or nutrified sawdust in bags or containers with holes or slits cut for fruiting sites at initiation time. Outdoors on logs laid on ground, or stumps. Often inadvertently cultivated, arising spontaneously from logs inoculated with other mushroom species. Because growth is so aggressive and adaptable, I suspect that turkey tail could learn to like eating many types of urban and ag wastes. Good fruitings from paper products have been reported. It may dominate on untreated sawdust.

Difficulty Level: Easy to grow indoors or out, though cloning tissue to liquid culture is challenging. Isolating on agar with tweezers is much easier. Cardboard culture is a viable option.

▲ Turkey tails growing under shade cloth in a greenhouse during winter, at Smugtown Mushrooms in Rochester, NY. OLGA TZOGAS

Growth Parameters: Moderately fast myceliation is followed by slow fruiting. Enjoy watching them grow for two months or so before harvesting. Fruiting temperature ranges from cool to warm.

Medicinal and Nutritional Properties: One of the best studied of all medicinal mushrooms, with a long history of traditional use in China, Mexico, Malaysia, and Australia. Potent, and broad spectrum in its effects. Some highlights: potently antitumor and helpful as an adjunct therapy to chemo and radiation; active against HIV and HPV; inhibits *E. coli, Candida albicans, Staphylococcus aureus*, and other human pathogens, while promoting the growth of probiotic *Bifidus* and *Lactobacillus*; supports liver, kidneys, and immune system, while also being anti-inflammatory and analgesic. Double extract is recommended. Recent findings of fat-soluble compounds confirm that my habit of throwing some in my chicken stock pot is a good idea.

Culinary Value and Use: While too tough to eat, its flavor is good and fits well in soups. Mushroom hunters commonly chew the fresh brackets. Some blend it to a pulp with water and add it to soups and sauces. One upscale bar in Vancouver infuses it in maple syrup and puts it in a cocktail with whiskey, lapsang tea tincture, lemon, and pineapple. In *The Wild Mushroom Cookbook*, Alison Gardner and Merry Winslow give a recipe for a vodka- or gin-based elixir with turkey tails, culinary herbs, and honey.

Comments: In addition to its medicinal properties, turkey tail's tenacious mycelium can bind various materials for living sculpture and functional materials, and the mushroom fibers left over from medicine preparation can be used to make fine paper. The mycelium produces a suite of powerful enzymes capable of degrading many toxic chemicals including PCPs, PAHs, synthetic dyes, pesticides, and pulp mill effluent.

APPENDIX 2
RESOURCES

GENERAL

All The Mushrooms (cultures, spawn, cloning service, great blog about fruiting room setup) Powell River, BC, allthemushrooms.com

Aloha Medicinals (culture bank, medicinals) Carson City, NV, alohamedicinals.com

American Type Culture Collection (ATCC) (immense culture bank) atcc.org

Amisacho (mycoremediation, ecological restoration, education, ecotourism, spawn, herbs) Lago Agrio, Ecuador, amisacho.com

Asheville Fungi (cultures, spawn, supplies, consultation, education) Asheville, NC, ashevillefungi.com

Bay Area Applied Mycology (mycoremediation, research, education, community lab, collective) Oakland, CA, bayareaappliedmycology.com

Catskill Fungi (medicinals, events, education) Big Indian, NY, catskillfungi.com

CoRenewal (bioremediation, research) USA and Lago Agrio, Ecuador and US, amazonmycorenewal.org

Cornell Mushroom Blog (info, free pdf handbooks, education) Ithaca, NY, blogs.cornell.edu/mushrooms

Cyberliber (online mycology library) cybertruffle.org.uk

DIY Fungi (cultivation info, education, remediation) Victoria, BC/Irvine, CA, diyfungi.blog

Everything Mushrooms (store with mushrooms, supplies, spawn, books, education) Knoxville, TN, everythingmushrooms.com

Far West Fungi (store with gourmet mushrooms, gifts, spawn, books, medicinals) San Francisco, CA, farwestfungi.com

FastFred's Media Cookbook (agar and LC recipes) shroomery.org/9393/FastFreds-Media-Cookbook

Female and Fungi (education, empowerment, community, blog, etc.) femaleandfungi.com

Field and Forest (spawn, supplies, info) Peshtigo, WI, fieldforest.net

Fresh Cap Mushrooms (spawn, supplies, info) Sherwood Park, AB, freshcapmushrooms.com

Fungaia Farm (spawn, cultures, education) Humboldt County, CA, fungaiafarm.com

Fungi Academy (education center, intentional community) Lake Atitlan, Guatemala, fungiacademy.com

Fungi Ally (cultures, spawn, education, supplies) Hadley, MA, fungially.com

Fungi for the People (cultivation and mycoremediation courses, research) Westfir, OR, fungiforthepeople.org

Fungi Magazine fungimag.com

Fungi Perfecti (cultures, spawn, supplies, books, education, medicinals) Olympia, WA, fungi.com

Hydrogen Peroxide Method Manual by Rush Wayne, PhD. mycomasters.com

Mushroom Growers' Newsletter (monthly periodical) mushroomcompany.com

The Mushroom Man (spawn, kits, education) Vancouver, BC, shroomstore.ca

Mushroom Mountain (cultures, spawn, supplies, education, schwag) Easley, SC, mushroommountain.com

Mycelium Emporium (huge selection of liquid cultures) Milford, ME, themyceliumemporium.com

The Mycelium Underground (community, education, events, social justice) Eastern US, themyceliumunderground.com

Mycoboutique (store with mushrooms, gear, education, supplies, spawn, books) Montreal, QC, mycoboutique.com

Mycolab Solutions (spawn, education, online mycology library) Occidental, CA, mycolabsolutions.com

Mycosupply (lab equipment, supplies) Pittsburgh, PA, mycosupply.com

MycoSymbiotics (cultures, information, education, Cordyceps) New Cumberland, PA, mycoshop.net

mycotek.org (cultivation forum)

mycotopia.net (cultivation forum)

Myco-Uprrhizal (cultures, spawn, myceliated textiles) Olympia, WA, mycouprrhizal.com

Mycoworks (fungal leather, mycelial material science, and design) San Francisco, CA, mycoworks.com

North American Mycological Association (non-profit, association of mycological societies) namyco.org

Northern Spalting (book, applied mycology info, fungal pigments, cultures, education) Corvallis, OR, northernspalting.com

North Spore (cultures, spawn, supplies) Portland, ME, northspore.com

Northwest Mycological Consulting (spawn, consulting, info) Corvallis, OR, 541-753-8198

Out-Grow (cultures, spawn, supplies) US, out-grow.com

Radical Mycology (info, education) radicalmycology.com

RR Video (*Let's Grow Mushrooms* instructional video series) mushroomvideos.com

shroomery.com (cultivation forum)

shroomology.org (cultivation forum)

Smugtown Mushrooms (spawn, medicinals, events, education) Rochester, NY, smugtownmushrooms.com

South African Gourmet Mushroom Academy (cultivation courses) Western Cape, South Africa, mushroomacademy.co.za

Terrestrial Fungi (cultures, spawn, consultation, education, mycoremediation) Warren, MI, terrestrialfungi.com

Uline (bags, poly tubing, gloves, sealers, packing, and shipping supplies, etc.) uline.com

Unicorn Bags (filter patch bags, cultivation supplies, info) Plano, TX, unicornbags.com

Violon et Champignon (cultures, spawn, education) Sainte-Lucie-des-Laurentides, QC, violonetchampignon.com

What the Fungus (internships, videos, Canadian distributor of Unicorn Bags) Summerland, BC, wtfmushrooms.com

Willoughby Arevalo (education, events, art, cultures) Vancouver, BC, mycelialconnections.net

ANNUAL MYCOLOGY GATHERINGS

International Fungi and Fiber Symposium—even-numbered years, various locations worldwide, mushroomsforcolor.com/SYMPOSIA.htm

MycoSymbiotics Mushroom and Art Festival—Pennsylvania, PA

Mycelial Mysteries—A Women's Mushroom Retreat, Wisconsin, midwestwomensherbal.com/mushrooms

NAMA Foray—(various locations in North America) namyco.org

New Moon Mycological Summit—(intersectional mycology gathering) Northeastern US, newmoonmycologysummit.org

NEMF Northeast Mycological Foray—Northeastern North America, nemf.org

Radical Mycology Convergence—(even-numbered years, US) radicalmycologyconvergence.com

Telluride Mushroom Festival—Telluride, CO, telluridemushroomfest.org

BIBLIOGRAPHY

BOOKS

Arora, David, *Mushrooms Demystified*. Berkeley, CA: Ten Speed Press, 1986.

Carluccio, Antonio, *Complete Mushroom Book*. London, UK: Quadrille Publishing, 2003.

Chang, S.T. and T.H. Quimio, eds., *Tropical Mushrooms: Biological Nature & Cultivation Methods: Volvariella, Pleurotus & Auricularia*. Hong Kong: The Chinese University Press, 1982.

Cotter, Tradd, *Organic Mushroom Farming and Mycoremediation: Simple to Advanced and Experimental Techniques for Indoor and Outdoor Cultivation*. White River Junction, VT: Chelsea Green Publishing, 2014.

Darwish, Leila, *Earth Repair: A Grassroots Guide to Healing Toxic and Damaged Landscapes*. Gabriola Island, BC: New Society Publishers, 2013.

Gardner, Alison and Merry Winslow, *The Wild Mushroom Cookbook: Recipes from Mendocino*. Mendocino, CA: The Barefoot Naturalist Press, 2015.

Holmgren, David, *Permaculture: Principles & Pathways Beyond Sustainability*. East Meon, UK: Permanent Publications, 2017.

Jadrnicek, Shawn and Stephanie Jadrnicek, *The Bio-Integrated Farm: A Revolutionary Permaculture-Based System Using Greenhouses, Ponds, Compost Piles, Aquaponics, Chickens, and More*. White River Junction, VT: Chelsea Green Publishing, 2016.

Klehm, Nance et al., *The Ground Rules: A Manual to Reconnect Soil and Soul.* Chicago, IL: Social Ecologies, 2016.

McCoy, Peter, *Radical Mycology: A Treatise on Seeing and Working With Fungi.* Portland, OR: Chthaeus Press, 2016.

Money, Nicholas P., *Mushroom.* Oxford, UK: Oxford University Press, 2011.

Oei, Peter, *Mushroom Cultivation: With Special Emphasis on Appropriate Techniques for Developing Countries.* Kerkwerve, The Netherlands: Backhuys Publishers, 1996.

Padilla-Brown, William, *Cordyceps Cultivation Handbook, Vol. 1: A Guide to Growing Cordyceps militaris.* New Cumberland, PA: Self-published, 2016.

Rice, Miriam C., *Mushrooms for Dyes, Paper, Pigments and Myco-Stix.* Mendocino, CA: Mushrooms for Color Press, 2007.

Roberts, Peter and Shelley Evans, *The Book of Fungi: A Life-Size Guide to Six Hundred Species from Around the World.* Chicago, IL: University of Chicago Press, 2011.

Rogers, Robert, *The Fungal Pharmacy: The Complete Guide to Medicinal Mushrooms and Lichens of North America.* Berkeley, CA: North Atlantic Books, 2011.

Rogers, Robert, *Mushroom Essences: Vibrational Healing from the Kingdom Fungi.* Berkeley, CA: North Atlantic Books, 2016.

Schaechter, Elio, *In the Company of Mushrooms: A Biologist's Tale.* Cambridge, MA: Harvard University Press, 1997.

Stamets, Paul and J.S. Chilton, *The Mushroom Cultivator: A Practical Guide to Growing Mushrooms at Home.* Olympia, WA: Agarikon Press, 1983.

Stamets, Paul, *Growing Gourmet and Medicinal Mushrooms.* Berkeley, CA: Ten Speed Press, 2000.

Stamets, Paul, *Mycelium Running: How Mushrooms Can Help Save the World.* Berkeley, CA: Ten Speed Press, 2005.

Steineck, Hellmut, *Mushrooms in the Garden.* Eureka, CA: Mad River Press, 1984.

Winkler, Daniel and Robert Rogers, Robert, *A Field Guide to Medicinal Mushrooms of North America.* Seattle, WA: Mushroaming Publications, 2018.

WEBSITES

Allison, Patricia, *The Principles of Permaculture: From Bill Mollison & David Holmgren.* Accessed 12/19/18. https://organicgrowersschool.org/wp-content/uploads/2013/06/THE-PRINCIPLES-OF-PERMACULTURE-Bill-Mollison-David-Holmgren-.pdf

FastFred, "FastFred's Media Cookbook." Accessed 8/12/2018. https://www.shroomery.org/forums/showflat.php/Number/3077006

Keith, Marc R., *Let's Grow Mushrooms.* Malo, WA, RR Video. Accessed 7/29/2018. www.mushroomvideos.com

Robinson, Dr. Seri, *DIY Spalting Pamphlets.* Accessed 10/29/18. https://www.northernspalting.com/spalting-info/diy-spalting-pamphlets/

Ross, Phil, "Mycotecture: Architecture Grown out of Mushrooms," talk at Parsons the New School of Design, filmed Tuesday, April 8, 2014. Accessed 9/16/2018. https://www.youtube.com/watch?v=7q5i9p0Yc3w

Tsing, Anna, "Unruly Edges: Mushrooms as Companion Species." Accessed 9/16/2018. http://tsingmushrooms.blogspot.com/2010/11/anna-tsing-anthropology-university-of.html

Many anonymous authors, many articles, www.shroomery.org. This is a website I use often.

JOURNAL ARTICLES

Garcia-Gonzalez, Rebeca et al., "Dietary inferences through dental microwear and isotope analyses of the Lower Magdalenian individual from El Mirón Cave (Cantabria, Spain)," *Journal of Archaeological Science*, 60, 28–38, August, 2015.

Heads, Sam W. et al. "The oldest fossil mushroom," PLOS ONE 12(6): e0178327, June 7, 2017. https://doi.org/10.1371/journal.pone.0178327. Accessed 7/24/18.

Mejía, Santiago Jaramillo and Edgardo Albertó, "Heat treatment of wheat straw by immersion in hot water decreases mushroom yield in *Pleurotus ostreatus*," *Revista Iberoamericana deMicología*, 30 (2):125–129, 2013.

INDEX

adaptation, and strain development, 89
aeration and maintenance, of LC, 79-80
agar, 27-28, 30, 36, 62, 81-88
agar plate to LC transfer, 77-78
agar slants, 88-89
agricultural wastes, as substrates, 110, 118-119
Agrocybe aegerita, 154. *See also* pioppino.
air filters, 103
airflow, 17, 29, 34, 38, 56-57
airport lids, 30, 56, 69, 70-71, 82
ale yeast, 123
American Type Culture Collection (ATCC), 68
Anaerobic Fermentation, 123
arbuscular mycorrhizal fungi (AMF), 14
ascomycetes, 6, 8
aseptic transfer space, 26-27
automation, and environmental control, 49

Bacillus subtilis, 62
bacteria, and fungi, 13-14
bacterial contamination, 62, 66, 81, 90, 93, 119
Bacti-Cinerator, 58
ballistospory, 10-11
basidia, 10
basidiomycetes, 6, 8-11

bathrooms, and fruitings, 42, 44, 53
battered and fried mushrooms, 142
biochar, 115
bioremediation, 2
blank substrates, 57
blewit mushrooms, 16, 115, 132
boletes, 14
brown-rotters, 16
Bryson, Kaitlin, 151
buckets, and cultivation, 44, 100, 103, 148, 167
bulk inoculations, 34, 53
Bunsen burners, 26, 30
business preparation, 24-25
button mushrooms, 1, 16, 112, 115

carbohydrates, and growth, 18-19
cardboard cultures, 88, 89, 111, 129, 168, 171
casing layers, 115
caterpillar fungus, 8, 156-157
Chencho, Pablo, 30
chestnut mushrooms, 166
chicken of the woods mushrooms, 16, 132
chlorine, in water, 18
classification, of mushrooms, 6-8
cloning mushrooms, 68
club fungi. *See* basidiomycetes.
coffee grounds, as substrates, 22, 34, 100, 112, 125, 167

cold water pasteurization, 123
commercial strains, 68, 168
community projects, 151-152
companion plants, 134
Complete LC (CLC), 73
compost decomposers, 16
compost mixes, as substrates, 112-113, 115
conditions, for growth, 17-19
containers, 98-104
contamination, 27, 28, 34, 37. 54-65, 81
cooking techniques, 139-142
Coprinus comatus, 155-156. *See also* shaggy mane.
Cordyceps militaris, 156-157. *See also* caterpillar fungus.
cultivation journal, 22
cultivation process, 20-24
cultivator, and contamination, 59
culture creation, 20
culture expansion, 21
cultures, defined, 66. *See also* spore prints.
culture storage methods, 88-89

decoctions, 145
decomposers, 15-16, 18, 155, 169, 170
dehumidifiers, 37
detergents, biodegradable, 124
dikaryons, 10, 68

Dog Food Agar (DFA), 84
double extracts, 145-146
dried mushrooms, 61, 141-142, 143, 145, 160, 163, 164, 165, 168

ecosystems, and mushrooms, 13-16, 148
ectomycorrhizae, 14
Egan, Mike, 4
elm oyster mushrooms, 16, 132, 162-163
enoki mushrooms, 38, 132, 157-158
environmental contamination, 59-60
environmental control, and fruiting, 46-51. *See also* airflow; fresh air exchanges; light; temperature.
ethics, 149-150

fiber wastes, as substrates, 111-112
field capacity, 105, 106, 112, 115
filter patch bags, 96, 98-99, 105
Flammulina velutipes, 157-158. *See also* enoki mushrooms.
flow hood, 26, 31
flushes, subsequent, 23, 41, 105, 115, 131-132, 135-136, 154, 158, 159, 161, 166
forest green molds, 63-64
fresh air exchanges (FAE), 36, 37, 38, 40, 41, 43, 49, 63, 115
frozen mushrooms, 144
fruiting, 19, 22, 44-52, 131-132
fungiculture, development of, 1-2
fusion, 10

Ganoderma lucidum, 159-160. *See also* reishi mushrooms.
garden oyster mushrooms, 162-163
gasteroid fungi, 11
genetic variability, 68
germination, 9-10
glass jars, 98
glove box, 28, 29
grain, types of, 90
grain cooking water LC (GCWLC), 73
grain or sawdust spawn to LC transfer, 78-79

grain spawn, 22, 27, 34, 36, 81, 89, 90-97
greenhouses, 51-52. *See also* mini-greenhouses.
growing medium, 20. *See also* agar; liquid culture.
gypsum, as additive, 73, 93, 105, 106, 112, 115

hardwood fuel pellets, as substrate, 104-105
harvesting, 23, 25, 37, 41, 61, 138, 148, 150, 156, 164, 168
HEPA filters, 29
Hericium species, 160-162. *See also* lion's mane mushrooms.
Honey LC (HLC), 73
humidifiers, 42-43, 44, 46-48, 49
humidity tents, 38, 51. *See also* mini-greenhouses.
hydration, and substrates, 105
hydrogen peroxide treatment, 124
hygrometers, 49
hyphal knots, 10
Hypsizygus tessulatus, 162-163. *See also* shimeji mushrooms.
Hypsizygus ulmarius, 162-163. *See also* elm oyster mushrooms.

incubation, 34-36. *See also* contamination.
incubators, 35-36
inoculation. *See also* bulk inoculation.
  of grain, 93, 95
  of logs, 104, 105, 127-129
  of nutrified sawdust with liquid culture, 107-108
  of pasteurized sawdust, 109
  of sawdust with grain or sawdust spawn, 108-109
inoculum, and contamination, 58
invasives, as substrates, 110

Keith, Marc R., 38, 106, 117, 121
kitchens, as labs, 30-31, 53, 56

labels, in cultivation, 22, 87, 129
lab infrastructure, 26-33
laminar flow hoods, 29
LC or spore syringe to LC transfer, 76-77, 80
Lentinula edodes, 163-165. *See also* shiitake mushrooms.
life cycle, of mushrooms, 8-11
light, and development, 36, 37, 43, 48
lignin, 2, 16, 18
lime, as additive, 112, 115, 122, 124
lion's mane mushrooms, 38, 99, 138, 160-162
liquid culture (LC), 20, 27, 36, 62, 69-81, 89. *See also* spore prints.
litter decomposers, 16
log culture, 64, 65, 126-132, 135, 154, 158, 164, 165. *See also* inoculation, of logs.

maitake mushrooms, 132
malt extract agar (MEA), 84
Malt Extract Dextrose LC (MDLC), 73
Malt Extract LC (MELC), 73
malt yeast agar (MYA), 84
Marthas. *See* mini-greenhouses.
Mason jars, 72, 98, 103
mating types, 11
medicinal preparations, of mushrooms, 145-146
medicinal properties, of mushrooms, 2, 8, 65, 97, 144-146. *See also mushrooms by name.*
metabolism, and growth, 17
mini-greenhouses, 42-43
mites, 61
molds, 8, 37, 58, 63-64, 81
monotubs, 40-41
morels, 8, 25
multispore germination, 75, 87
mushroom beds, 132-135
Mycality Mushrooms, 3
mycelial materials, 150-151
myceliated grain, 97
myceliation, 10

mycelium, background to, 6, 9-10.
    See also cultivation process.
mycoarts, 150-151
mycopermaculture, 147-150
mycoremediation, 89, 150, 152
mycorrhizal species, 13, 14

nameko mushrooms, 132, 135, 138,
    165-166
nutrients, and growth, 16, 18-19, 97,
    110-111, 123
nutrified sawdust, and fruiting, 35, 62,
    98, 106

online community, 30, 40, 42, 56, 66, 70
outdoor fruiting, 51-52, 53
oven tek, 26, 30
oversupplementation, 62
oyster mushrooms, 15, 18-19, 37, 38,
    41, 44, 97, 99, 114, 124. See also
    elm oyster mushrooms; garden
    oyster mushrooms; phoenix oyster
    mushrooms.

paddy straw mushrooms, 16
palm inoculator, 127-128
papermaking, 151
paper products, as substrates, 111, 114
parasitic species, 13, 14-15, 54
parasol mushrooms, 16, 132
pasteurization, 118-123
pasteurized plain sawdust, 109
permaculture principles, 147-150
pests, and contamination, 60-62
Petri dishes, 20, 77, 81, 82
pH, of fruiting substrates, 112
pH treatment, 124
phoenix oyster mushrooms, 89, 166-168
Pholiota nameko, 165-166. See also
    nameko mushrooms.
pickled mushrooms, 143-144
pins. See primordia.
pioppino, 115, 132, 135, 138, 154
plastic bags, 98-99, 103
pleasing fungus beetles, 60-61

Pleurotus citrinopileatus, 166-168
Pleurotus djamor group, 166-168
Pleurotus eryngii, 166-168
Pleurotus ostreatus, 166-168. See also
    oyster mushrooms.
Pleurotus populinus, 166-168
Pleurotus pulmonarius, 166-168. See
    also phoenix oyster mushrooms.
polyethylene bags, 99, 103, 109
polypropylene bags, 98-99, 103
portobello mushrooms, 112
powdered mushrooms, 143
pressure canners, 116-117, 118-119
primary decomposers, 15-16
primordia, 10, 22

racks, for fruiting, 44
recipes
    agar media, 84
    compost-like blends, 115
    double extract tincture, 146
    liquid media, 73
    sawdust fruiting blocks, 106
reishi mushrooms, 41, 132, 135, 159-160
relative humidity, 36, 37, 38, 49
rest phase, 23
Rice, Miriam C., 151
roasting/grilling mushrooms, 142
Robinson, Seri, 151
rocket stoves, 121
rodents, 61-62
Rosen, Carmen, 151
Ross, Phil, 150

Sabouraud's Dextrose Broth (SabDex), 73
sac fungi. See ascomycetes.
saprotrophic species, 13, 15-16, 154,
    160-165, 169-172. See also oyster
    mushrooms.
sautéed mushrooms, 139-141
sawdust, as substrate, 34, 44, 52, 100,
    104, 105, 109, 117, 126. See also nutri-
    fied sawdust fruiting blocks.
sawdust spawn method, 78-79, 96,
    108-109, 114, 127-129, 132

secondary decomposers, 15
secondary metabolites, 17
self-healing injection ports (SHIP), 56,
    69, 71, 99
shaggy ink cap mushrooms, 155-156
shaggy mane mushrooms, 16, 115,
    155-156
shiitake mushrooms, 19, 44, 68, 99, 100,
    112, 126, 131-132, 138, 144, 163-165
shimeji mushrooms, 16, 162-163
shotgun fruiting chambers, 38, 40
shower stalls, and fruitings, 42
solar pasteurization, 122-123
soups, 141-142
spawn expansion, 22
spawn generation, 22
spawn run, 129, 131
spent substrate, 23, 52
spore prints, 66-68, 87-88, 89
spores, 8-11
spore slurries, 134
spore syringes, 75-77
spraying or broadcasting, 134
steam pasteurization, 121-122
sterilization, 116-117
stewardship, 148, 151
stick spawn, 96
still air box, 26, 28-30
stir bars and plates, 80
stock, from mushrooms, 142
storage, of mushrooms, 138
strains, 10, 19, 20, 44, 68, 75, 81, 87, 89,
    149, 158, 160, 161
straw, as substrate, 23, 34, 52, 53, 64, 68,
    89, 99, 109-110, 123, 124
strawsages, 99
Stropharia rugoso-annulata, 169-171.
    See also wine cap mushrooms.
stump cultivation, 132
subculturing (subbing), 87
submerged fermentation. See liquid
    culture.
substrates, 15-16, 17, 18-19, 35, 52.
    See also bulk inoculations; fruiting;
    pasteurization; and by name.

and contamination, 62, 63-64
recycling of, 135-136
treatment of, 57, 125
types of, 104-115
super pasteurization, 117
supplies, 32-34
symbiotic fungi, 13-14

taxonomy, of mushrooms, 6-8
temperature, and growth, 19, 34, 36, 37, 46, 68
tertiary decomposers, 15, 16
thermometers, 49
thermophilic composting, 112-113
timers, 49
tinctures, 145-146
tissue culturing to agar, 85-87
tissue culturing to liquid culture, 73-75
tools, 32-34, 57-58
Trametes versicolor, 65, 171-172. See also turkey tail mushrooms.
trash production. See waste reduction.
trays, 102
Trichoderma. See forest green molds.
troubleshooting, 51
truffles, 8, 11, 14
turkey tail mushrooms, 132, 171-172
Tyvek, 28, 60, 99, 103, 106

ultraviolet light, 30
urban waste, as substrates, 110-112

ventilation, 44, 48. See also airflow.
vessels, for fruiting, 38, 51, 102-103. See also by name.
viral infections, 65

washing machine treatment, 124
waste reduction, 57, 100, 103, 114, 143, 148, 149
water, and growth, 17-18
water bath pasteurization, 119-121, 122
Wayne, Rush, 124
weed fungi, 64-65
weeds, as substrates, 110
wet-spot, 62, 93
white-rotters, 16
wild mushrooms, 16, 37, 68, 89, 153, 156, 157, 160, 163, 169-171
wine cap mushrooms, 25, 110, 115, 132, 133, 138, 169-171
woodchips, as substrate, 64, 122, 123, 132-133, 134, 154, 162, 167, 169
woodchip spawn, 53, 135
wood-rotters, 15-16, 19
work flow, 52, 59
workspace cleaning, 26-27, 29-30, 31, 34, 37, 59-60, 103

yeasts, 8, 64

ABOUT THE AUTHOR AND ILLUSTRATOR

Willoughby Arevalo. ISABELLE KIROUAC

WILLOUGHBY AREVALO is passionate about the ecology of fungi, the ways they shape our world, and the ways we shape theirs. His lifelong friendship with fleshy fungi has led him down a mycelial pathway—from his start in field identification and mushroom hunting, branching into cuisine, DIY cultivation, farming, education, and eco-arts. In his over 30 years of self-motivated inquiry and intimate lived experience with fungi, he has spent the last decade sharing mycology with people in communities across North America. This has manifested in collaborations with the Guapamacátaro Art and Ecology Center in Mexico; Isabelle Kirouac; Homestead Junction; Swallowtail Culinary Adventures; Lynn Canyon Ecology Centre; the Italian Cultural Centre; Sunshine Coast Mushroom Festival; Robert Rogers; North Van Arts and the Vancouver Mycological Society in Canada; The Mycelium Underground and the New Moon Mycology Summit; CoRenewal's Art and Science of Mycorenewal; Telluride Mushroom Festival; Radical Mycology Convergences; tours, and more in the US. He is currently a mycoartist in residency at Mountainside Secondary School in North Vancouver, and Still Moon Arts Society's Alder Eco Arts Hub. When not busy with all that and caring for his kid, Uma, he works part-time on an organic vegetable farm. Originally from Arcata, California (traditional Wiyot and Yurok Territory), he lives as a guest on unceded Coast Salish Territory in Vancouver, Canada. He can be reached through his website, mycelialconnections.net.

CARMEN ELISABETH is an Oakland, CA-based artist who loves to tinker in all sorts of media. She seeks to tell stories with her work, whether it's large-scale live painting at an event or an intimate suitcase puppet show. She likes to make little intricate things and big wild things, to curate large-scale visual experiences, and to create interactive art and performance. She also particularly likes collaborating with other artists and facilitating community art projects—to explore art as a vibrant language of expression, presenting visions of what we hope and dream for in our communities and our world.

Carmen Elisabeth. FALLYN MCLEOD

A NOTE ABOUT THE PUBLISHER

New Society Publishers is an activist, solutions-oriented publisher focused on publishing books for a world of change. Our books offer tips, tools, and insights from leading experts in sustainable building, homesteading, climate change, environment, conscientious commerce, renewable energy, and more — positive solutions for troubled times.

We're proud to hold to the highest environmental and social standards of any publisher in North America. This is why some of our books might cost a little more. We think it's worth it!

- We print all our books in North America, never overseas
- All our books are printed on 100% post-consumer recycled paper, processed chlorine free, with low-VOC vegetable-based inks (since 2002)
- Our corporate structure is an innovative employee shareholder agreement, so we're one-third employee-owned (since 2015)
- We're carbon-neutral (since 2006)
- We're certified as a B Corporation (since 2016)

At New Society Publishers, we care deeply about what we publish—but also about how we do business.

New Society Publishers
ENVIRONMENTAL BENEFITS STATEMENT

For every 5,000 books printed, New Society saves the following resources:[1]

19	Trees
1,752	Pounds of Solid Waste
1,927	Gallons of Water
2,514	Kilowatt Hours of Electricity
3,184	Pounds of Greenhouse Gases
14	Pounds of HAPs, VOCs, and AOX Combined
5	Cubic Yards of Landfill Space

[1] Environmental benefits are calculated based on research done by the Environmental Defense Fund and other members of the Paper Task Force who study the environmental impacts of the paper industry.